Single-Molecule Super-Resolution Optical Microscopy

Partha Pratim Mondal • Samuel Hess

Single-Molecule Super-Resolution Optical Microscopy

Springer

Partha Pratim Mondal
Instrumentation and Applied Physics
Indian Institute of Science
Bangalore, Karnataka, India

Samuel Hess
Department of Physics & Astronomy
University of Maine
Orono, ME, USA

ISBN 978-3-032-13441-7 ISBN 978-3-032-13442-4 (eBook)
https://doi.org/10.1007/978-3-032-13442-4

This Springer imprint is published by the registered company Springer Nature Switzerland AG
The registered company address is: Gewerbestrasse 11, 6330 Cham, Switzerland

Resolving smallest feature (single molecules) beyond the limit set by natural laws is the biggest innovation of this century. The application are limit-less.

—*Partha Pratim Mondal, Self-Thought*

Single molecule localization microscopy is a powerful tool for understanding biological and chemical systems at the nanoscale. Yet, it is also very accessible to non-experts, who through this book we hope will be encouraged to learn more and become the experts of the future.

—*Sam Hess, Super-Resolution Inventor*

Dedicated to my mentors (Prof. Richard Blake, Prof. Daniel Prober, Prof. Watt Webb, and Dr. Joshua Zimmerberg), my daughters (Adelaide and Sarah Kate Hess), my wife (Prof. Julie Gosse), and my parents (Dr. Anne L. Hess and Prof. Thomas Hess).

Dedicated to my mentors (Prof. Kanhirodan Rajan, Prof. Alberto Diaspro, Prof. Peter So, and Prof. Richard Gilbert), my daughters (Manasvini and Nakshatra), and my parents (Gita Mandal and Kartik C. Mandal).

Introduction

Until 1990s, it was believed that resolving features beyond the classical limit governed by physical laws (diffraction limit) was not possible. So, the very ability to circumvent this limit is of fundamental importance and quite fascinating for understanding our surroundings (both living and non-living). Until recently, harvesting information has been largely facilitated by diffraction-limited systems which limit us to a lower resolution. This means that our understanding at the fundamental level (single molecule level) is limited and mostly governed by indirect measurements/experiments. So, techniques that allow direct visualization of the fundamental functional entities of life (single biomolecule such as, single protein/DNA) in real-time are of greatest value to understand the world around us and the way it behaves. This book is intended to discuss powerful microscopy and imaging techniques that are capable of resolving features much below the diffraction limit of light with a strong emphasis on single molecule physics. Such optical techniques are key to reveal the nanoscopic world and some of its smallest functional entities (single molecules).

The last decade has seen the emergence of a new kind of optical microscopy that is not limited by the classical diffraction limit. This is amazing considering the fact that such powerful microscopes can resolve features and cause a PArticle Resolution shift (PAR-shift) toward the size of actual single molecules, thereby bringing a plethora of insight into the functioning of biological processes critical to living organisms. In general, light is known to diffract, and this places a fundamental limit on the resolvability of sub-diffraction features. As a result, any point source visualized through lens-based systems (microscope or other imaging systems) appears not as a point but an airy-disc pattern with a large central lobe associated with concentric side lobes. This is called the point spread function (PSF). This has been interpreted to mean that, one cannot resolve features smaller than half the size of central lobe which is approximately half the wavelength of light ($\sim \lambda/2$). Thus, it is imperative to understand light in its entirety and its interaction with single molecules to have a better understanding of optical resolution and ways to circumvent it.

Light microscopy is attractive largely due to its noninvasive nature and ability to resolve objects at small length scales. The best way to comprehend optical microscopy is to understand light propagation and its interaction with matter/single molecules. Light is explored by four core theories Ray Optics, Wave Optics, Electromagnetic Optics, and Quantum Optics that cast light as rays, waves, electromagnetic wave, or discrete quantum entity (photon), respectively. Other intermediaries such as Polarization Optics, Fourier Optics, and Plasmonics are essentially derivatives of these four basic theories. Super-resolution optical microscopy is one such system that makes use of all these forms (ray/wave/ photon) of light. In practice, realizing an optical microscope requires ray optics for alignment and maneuvering light, wave optics for understanding diffraction of light, electromagnetic optics for realizing the effect of electric field and its polarization, and quantum optics to decipher events occurring at discrete energy level or single photon level (e.g., photon emission from a single molecule). All these theories (of which some are underexplored yet) empower existing microscopes and paves the way for building futuristic and powerful super-resolution microscopes.

Super-resolution optical microscopy as we know today has the remarkable ability to defy the classical diffraction limit. These are indeed a family of super-resolution microscopes that are extensively used in biology and are beginning to penetrate other research disciplines (physical and material science, chemical science, and biomedical engineering). These specialized microscopy techniques can be classified into following categories, largely based on the physical concept: (1) SMLM (fPALM/PALM/STORM/PAINT), (2) GSDIM/RESOLFT/STED, and (3) SIM/TIRFM. While SMLM techniques largely depend on the wave optics (diffraction of light) and the number of quanta/photons detected per molecule (with computational processing of single molecule data), the techniques based on RESOLFT work on the principle of stimulated emission depletion of molecules (electromagnetic optics). However, SIM and related techniques are phase based generation of structured light (uses wave-nature of light which is a sub-domain of wave optics). Realization of all these techniques requires ray optics and hence a combination of ray optics, wave optics, and electromagnetic optics is central to super-resolution microscopy.

Another classification could be based on the fluorescent probes and their properties, i.e., (1) Photoactivable Proteins, (2) Phtotswitchable Dyes, (3) Regular Fluorescent Dyes. These specialized proteins and dyes have the special ability to switch between two states (ON/OFF) when exposed to an appropriate wavelength of light. This is crucial since this enables switching a randomly selected small fraction of these fluorophores at any time, which can then be excited and identified as single molecules. Within two decades of the first discovery of a switchable fluorophore (GFP in the year 1996), a host of such fluorophores (Dendra2, mEos, Dronpa, pamCherry, TPA-DHPy, etc.) are now available. We expect this to continue and to aid holistic development of super-resolution microscopy. Of late, the technique is gaining popularity in natural sciences (ranging from cell biophysics to clinical biology) and physical sciences (from single molecule physics to photochemistry), and now reaching the engineering sciences (materials engineering, nanotechnology, and chemical engineering) as well. With the advent of super-resolution microscopy, the developments in diverse research disciplines are expected to accelerate and give birth to several new interdisciplinary fields in near future.

Scope

This book on Single molecule Super-resolution Optical Microscopy introduces the fundamentals of light, single molecules physics, and recent key developments in the field. The book is coherent in handling the subjects and takes the reader from the basics of light to the advanced single molecule imaging. The following specialized topics and research areas are discussed in this book:

1. General imaging optics and behavior of light as it propagates through optical elements.
2. The design of key optical systems (beam-expander, magnifiers, relay-systems, 4f system, etc.) essential for realizing a complete imaging/microscopy systems.
3. Wave nature of light and the important aspect related to the diffraction of light for realizing optical microscopy.
4. The polarization of light, and its interaction with single molecules (treated as dipoles) that are central to single molecule physics.
5. The quantum aspect of light and its integration with optical microscopy.
6. The instrumentation behind the illumination and detection sub-system, and ultra-sensitive single photon detectors necessary for single molecule detection.
7. The photophysics of single molecules, emission characteristics, and the chemistry behind it.
8. Different classes of single molecules based on their chemistry such as photoactivable, photoswitachable, and photoshiftable molecules.
9. Key super-resolution microscopy techniques: fPALM/PALM/STORM/STED/SIM and variants.
10. The application of super-resolution microscopy in biological and physical sciences.

Single molecule physics/biology has been a revolution in the modern era. This is an emerging discipline at the interface of physical, biological, chemical, and engineering sciences. In the next decade, this may have implications in all aspects of nature sciences, and it is poised to make new discoveries.

Unique Features

Unlike the existing literature on the subject, this present book has the following unique features:

1. The book begins with the basics of light (ray/wave/electromagnetic/quantum optics) and gradually takes the reader to the most advanced techniques ever invented, e.g., single molecule-based super-resolution microscopy/imaging.

2. The book does not assume any prerequisites which makes it easy and friendly for students from a wide research disciplines (natural sciences, engineering, and medical science).
3. The book is introductory, conceptual, and mostly non-mathematical in nature.
4. The book features special topic in quantum optics which is envisioned as the future of optical microscopy and super-resolution microscopy in particular.
5. The book serves the dual purpose of serving both the beginners who wish to enter the field and experts who wish to make new advance.
6. The book may serve as an asset for researchers (biologists, physicist, pharmacologists, and clinical physicist) working in interdisciplinary research disciplines who wish to benefit from super-resolution microscopy.

Intended Audience

This book is an introductory and cohesive representation of this expanding field of super-resolution microscopy. The book may serve as:

1. An introductory text for students in biological, physical, and engineering sciences who want to enter the field.
2. A simple textbook for self-study and a valuable reference for optical microscopy and single molecule imaging.
3. A valuable text for researchers, graduate students, professionals, scientists, and engineers both in academia and industry.
4. A reference connecting classical and quantum optics to microscopy/imaging.
5. A useful guide to experts working in this emerging field at the interface of natural sciences, engineering, and medical science.

The book is written in a way that we hope suits everyone's needs from undergraduates to experts and does not necessarily require strong background in physics or optics or mathematics. The book is on the necessary fundamentals, begins with an introduction to imaging optics with special emphasis on diffraction of light, and takes the reader to the advances in single molecule microscopy/imaging in a gradual manner. Consciously, the book is kept simple and extreme mathematics is avoided, in fact certain aspects of the book are backed by logical arguments and simple reasoning. The book is self-contained and complete, so the reader may not require additional references. Further, we have taken-up few examples/case studies that are specific to certain applications and related to specialized versions of single molecule microscopy. For serious readers who wish for mathematical rigor, references are incorporated at the end of each chapter to further explore. To our knowledge, this will be the first book that integrates fundamental concepts of optical physics with super-resolution microscopy.

Book Organization

This book is intended to be a single holistic textbook that integrates fundamental optics (both classical and quantum optics) to the advanced concepts in super-resolution microscopy. The book is futuristic due to the inclusion of quantum aspects of light, which is expected to play a vital role in the future development of single molecule imaging. The chapters are organized in two major parts: Imaging Optics and Super-resolution Microscopy. The first part deals with fundaments of imaging optics that prepares the reader for basic understanding of light and functioning of optical components, while the second part deals with advanced microscopy techniques/optical designs that enable imaging beyond the classical diffraction limit of light with single molecule precision. The details are as shown below:

Fundamentals of Imaging Optics

1. Introduction to Super-Resolution and Its Impact
2. Basics of Ray Optics for Optical Microscopy
3. Diffraction of Light and Classical Resolution Limit
4. Electromagnetic Optics, Light-Matter Interaction and Optical Microscopy
5. Quantum Optics for Precision Microscopy

Single Molecule Physics and Super-Resolution Microscopy

6. Introduction to Molecular Physics and Fluorescence Spectroscopy
7. The Photophysics/Photochemistry of Single Molecules Probes
8. The Signature of Single Molecule
9. Optical Design and Instrumentation of Single Molecule Optical Microscope (fPALM/PALM/STORM)
10. Biotechniques and Cell Biology Protocols for Single Molecule Biology
11. Stimulated Emission Depletion Microscopy (STED) and Its Single Molecule Analogues
12. Selected Topics in Single Molecule Biophysics

Representative Courses

The chapters can be combined to represent a couple of single semester courses: (1) Fundamentals of Imaging Optics and, (2) Single Molecule Physics and Super-Resolution Microscopy.

Fundamentals of Imaging Optics

A course in Imaging Optics comprises classical theory of light. This includes the basic principles of light and the laws governing its propagation. The study of optical elements used in optical microscopy and its effects such as reflection from a mirror, refraction/transmission from a beam-splitter, beam-expander for beam-expansion, objective lens for tight focusing, eyepiece, etc. This is followed by the conceptual understanding of diffraction and interference of light that play key roles in optical microscopy such as 4pi microscopy, single molecule interference microscopy, and multi-focus microscopy. Then, electromagnetic optics is discussed to understand polarization effects of light and its interaction with electric dipole (fluorophore). Finally, quantum optics that allows the use of quantum aspects of light to enhance imaging and empower optical imaging with new capabilities. Based on these specialized topics in imaging optics, we will also discuss microscopy and optical devices used in medical optics and optical imaging.

Single Molecule Physics and Super-Resolution Microscopy

A course in Single Molecule Physics and Super-Resolution Optical Microscopy may begin with the fundamentals of molecular physics with a special emphasis on single molecule. This may be followed by a brief discussion on the chemistry of dyes, fluorescent proteins, and photoactivable fluorophores. Subsequent topics related to optical design of super-resolution microscopes and imaging systems may be taken up. Sample preparation and selectivity of specific proteins in living cells/organisms plays critical role and so this needs a detailed discussion. Super-resolution techniques can be classified based on the illumination scheme, i.e., Widefield/TIRF/HILO, which forms the base for a specific application or study in hand. This needs special attention and depends on the application in hand. Subsequently, three major super-resolution techniques and the physics behind them may be discussed. For example, single molecule SMLM techniques (fPALM, PALM, STORM, and STED microscopy) depend on the fluorescent state of single molecule (ON/OFF), whereas STED and SIM are based on stimulated emission depletion and patterned light illumination, respectively. Finally, selected topics in single molecule super-resolution techniques that include hybrid techniques (integrating two or more microscopy techniques) and emerging new techniques along with their applications may also be discussed.

Acknowledgements

We acknowledge all the faculties, experts, and researchers who have contributed directly or indirectly toward this book. We also acknowledge collaborators and friends for their valuable suggestions and comments. I am in debt to my parent institution (Indian Institute of Science, Bangalore, India) for providing me with an environment conductive to much-needed academic freedom that allowed me to take up this project along with teaching and research.

I drew on numerous science and engineering books and journal articles that influenced the topics covered in this book. Indeed, many figures and illustrations are borrowed, for which appropriate permissions were obtained, and references are cited.

I am grateful to my friend and collaborator, Prof. Samuel T. Hess (University of Maine, Orono, Maine, USA), for agreeing to partner with me on this important book project and for all the friendly discussion on various aspects of light and microscopy. I am grateful to Prof. Malvin Carl Teich (Professor Emeritus, Boston University, MA, USA, and Columbia University, NY, USA) for the necessary references on Quantum Optics. Many thanks to my mentors, Prof. Alberto Diaspro (IIT, Genova, and University of Genova, Italy), Prof. Peter T. C. So (Massachusetts Institute of Technology, MA, USA), and Prof. Richard Gilbert (Northeastern University, MA, USA) for introducing me to optical imaging. Special thanks to Prof. Alberto Diaspro (IIT, Genova, and University of Genova, Italy) and his lab for introducing me to the wonderful field of light microscopy, and for the continued support over the years. I am grateful to have great collaborations, Dr. Franchesca Cella (University of Pisa, Italy), Dr. Subhra Mandal (University of Nebraska, Lincoln, USA), Prof. Utpal Nath (Indian Institute of Science, Bangalore, India), Prof. Chandrasekhar Seelamantulla (Indian Institute of Science, Bangalore, India), Dr. Nikhil Gandasi (Indian Institute of Science, Bangalore, India), Prof. Upendra Nongthomba (Indian Institute of Science, Bangalore, India), Dr. Abhishek Kumar (Marine Biological Laboratory, Woods Hole, MA, USA), Dr. Suheshkumar Singh (IISER, Trivandrum, India), and Dr. Matthew Parent (Marine Biological Laboratory, Woods Hole, MA, USA), who have helped me shape my research that has a lot of impact on this book.

The book is modeled on the course (CCT302: Single Molecule Imaging and Cryo-electron Microscopy) that I teach at my parent institution (Indian Institute of Science, Bangalore, India) and the class notes that I have prepared over the years. The active participation of students in the course has helped me shape the chapters. I am grateful to my research students (Prashant Kumar, Prakash Joshi, Jigmi Basumatary, Aravinth S, Neptune Baro, Neeraj Pant, and Souvik Saha) and postdoctoral researchers (Dr. Jiby M. Varghese) for going through the chapters and providing me feedback from a student's perspective. Many of them have gone through the chapters many times and provided me suggestions for minor corrections. Many thanks to numerous colleagues and fellow faculties from the parent institute (Indian Institute of Science, Bangalore) for collaborative support.

Dr. Hess would like to thank C.T. Hess, Anne Hess, Julie Gosse, George Patterson, Joshua Zimmerberg, Dean Astumian, Dominique Bourgeois, Rosemary Smith, Scott Collins, Michael Mason, George Bernhardt, Joerg Bewersdorf, Travis Gould, Watt Webb, Michael Mlodzianoski, Nikki Curthoys, Neil Comins, and Christoph Cremer for useful discussions, colleagues at U. Maine, Yale, NIH, and Cornell, as well as current and former Hess lab members for ideas and support.

A heartfelt thanks to Zachary Evenson (Editor, Springer-Nature), Aishwarya Srinivasan (Springer-Nature), Antje Endemann (Springer-Nature), Srimathi V (Springer-Nature) and others at Springer-Nature publishing for giving us the opportunity to write this book.

Indian Institute of Science, Bangalore, India Partha Pratim Mondal
University of Maine, Orono, ME, USA Samuel Hess

Contents

List of Figures

List of Tables

Introduction to Super-Resolution and Its Impact 1

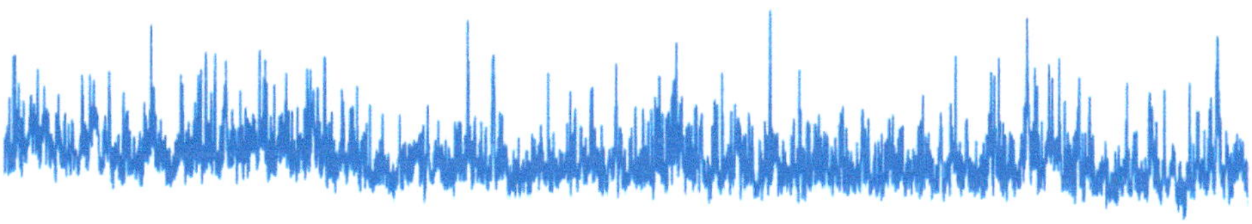

Physical laws set natural limits on how much nature wishes to reveal itself. This is appropriate for light where diffraction of light sets a limit on the resolvable distance. Another limitation is associated with the size of the smallest functional unit of matter (i.e., single molecule) that sets the length scale for investigation. Such limitations on both the probe (here, light) and the matter hold scientific progress, and our understanding of natural processes at the fundamental scale (single molecule). This is true for the field of optical microscopy and related sciences, where the diffraction limit on the resolution has ceased its progress for centuries. The years 1990s and 2000s saw the emergence of three distinct classes of microscopy techniques (STED, SMLM, and SIM) that surpassed the classical limit. Among them, single molecule localization microscopy techniques (fPALM, PALM, and STORM) could reveal the true nature of single molecules. These techniques, for the first time, revealed the nanoscopic world of life processes, including how proteins are manufactured, their interactions, biological functions, and so on. These and many other fundamental processes occurring at single molecule length scales have profound implications in wide research disciplines ranging from natural to engineering sciences.

Often, realizing capabilities better than what nature dictates has taken human endeavors to a new level. In this respect, the emergence of super-resolution has overcome the diffraction limit (a core property of classical light) that opened up the nanoscopic world of living objects. This has not only pushed the boundaries of light itself but has infused new life into many branches of natural sciences which were stuck for a very long time.

P. P. Mondal, S. Hess, *Single-Molecule Super-Resolution Optical Microscopy*,
https://doi.org/10.1007/978-3-032-13442-4_1

Among them, the most affected is the biological sciences where resolution has caused a blurry/fuzzy world of biological processes.

Just after the discovery of the resolution limit by Abbe and Rayleigh in early 1900, it convinced the scientific community that diffraction is an inherent property of light, and it can't be surpassed beyond the classical limit $\sim \lambda/2$ [1–3]. However, progressing scientific advances demanded better resolution in order to understand the physical world at large and life sciences at specific. Although diffraction is a natural phenomena and brings out the inherent nature of light, this is also a stumbling block for further scientific investigation into the physical world that we live in. For example, understanding subcellular processes and the function of specific proteins has been a long-term problem for humanity. The same goes for understanding plants at the single molecule level, with questions such as which proteins/biomolecules are involved in photosynthesis, how they carry out specific tasks, how much time it takes for specific task, and what determines their efficiency from a single molecule perspective. As far as prokaryotic and eukaryotic life forms are concerned, how different biological processes are carried out at the single molecule level remains an area of investigation. These biological processes can be from the perspective of a simple task of food intake to resisting infection by viruses and bacteria. These questions are essential and must be understood at the very molecular level to thwart pandemics (Covid-19) and epidemics that are known to wipe out civilizations and are a constant risk to humanity. All these reasons and many others suggest the need for an investigative tool that can visualize and interrogate processes and functions at sub-diffraction length scales.

Barring others, the actual quest for super-resolution and ways to surpass the diffraction limit started in 1952 by Italian Physicist and Philosopher Toraldo di Francia. He proposed pupil-plane filtering (a set of concentric amplitude and/or phase filters at the pupil of the imaging lens) to increase system resolution [4]. This is the first time the quest for super-resolution has been proposed and demonstrated, focusing on a spot smaller than the Abbe-Rayleigh diffraction limit. The quest for super-resolution continued but was briefly forgotten with the advances in electron microscopy that produced much better resolution (<1 nm) and without the need for labeling or specialized optical arrangements. In the subsequent years, another technique became popular: Near-field scanning optical microscopy (NSOM). Although the technique is largely near field, nonetheless, it could produce a high-resolution map of the specimen surface morphology. In NSOM, the diffraction limit is overcome by using a sub-wavelength probe (typically a tapered optical fiber or a metal-coated tip) and exploits near-surface fields (evanescent waves), which are not confined in the same way by diffraction. In the process, the probe is scanned across the sample surface, and its interaction with the near-surface evanescent waves allows the detection of features smaller than the diffraction limit of light ($\sim \lambda/2$). Since the technique is generally limited to the specimen surface, it could only generate topological images of the surface, i.e., it has a limited depth of field. This is not fully satisfying what was actually desired by the scientific community i.e., depth imaging, touch-free interrogation, and less stringent conditions for live specimens.

1.1 Super-Resolution Microscopy: A Brief History

The race for super-resolution started to take shape in the early years of 1980, when techniques such as standing wave illumination using opposing objectives in confocal (4Pi) or widefield (I5M) microscopy were proposed [5–8]. In the early 1990s, another technique called "STimulated Emission Depletion (STED) microscopy" took shape that demonstrated multifold improvement in lateral resolution [9]. Subsequent years have seen the development of two major super-resolution microscopy techniques: structured illumination microscopy (SIM) and single molecule localization microscopy (SMLM) [10–17]. However, it took a few more years for the technology to mature and adapt to a widespread research fields. The technology requires highly sensitive detectors, light sources/lasers, computers with high processing powers, and better photoswitchable/photoactivable fluorophores. At this time point (~2006), two major super-resolution microscopy techniques were established: SIM, and SMLM. Of these, SMLM techniques have gained much prominence, which can be attributed to its simple implementation, affordability, and the ability to identify single molecules. As we stand today, three major SMLM techniques i.e, fPALM, PALM and STORM are in widespread use.

1.2 Major Advances in Single Molecule Based Super-Resolution Microscopy

In this book, our focus will be on SMLM techniques and related key variants. These techniques have progressed primarily to understand biological processes spanning entire biological sciences and even spread to related fields such as, biomedical sciences, biotechnology, biochemistry and drug discovery. Some of the advanced features of SMLM are (i) Improving spatial resolution, (ii) Volume imaging, (iii) Imaging at large penetration depths, and overall high spatio-temporal resolution. From a pure application viewpoint, the technique is used to decipher intra-cellular processes during viral infection and disease biology, apart from understanding fundamental biological processes.

Regarding point (i), a handful of microscopy techniques have shown potential. Achieving three-dimensional resolution (both lateral and axial) was proposed by the Samuel Hess group in 2008 and still remains an important technique (BP-fPALM) for 3D cell imaging [18]. The technique has shown an impressive spatial resolution of $30 \times 30 \times 75$ along X, Y, Z axis at a depth of several micrometers. Another method by Xiaowei Zhuang group demonstrated 3D nanoscale resolution imaging using optical astigmatism [19]. However, the technique that boosts of sub-diffraction optical resolution in all three dimensions is 3D-SOFI [20]. The technique uses temporal fluctuations (caused by fluorescence blinking) and has demonstrated an improvement in spatial resolution by 5-fold. In similar lines, Pavani et al. demonstrated 3D SMLM using a double helix PSF [21]. On this method, the PSF is such that its angular orientation rotates with the Z-position of the emitter and can image at large depths of $\sim 2\,\mu$m. A different approach is demonstrated by Hajj et al., where the concept of multifocal imaging is conjugated with PALM [22]. The technique could produce 9 different focal planes at a time and is shown to image at a depth of 4 μm with a localization precision of 20 nm laterally and 50 nm axially.

As far as volume imaging in SMLM is concerned, biplane fPALM (BP-fPALM), 3D STORM, tilted lightsheet SRM, and $scanSMLM$ are some of the promising techniques. For example, a paired tilted light sheet is used to illuminate and excite single molecules in a selective plane, and the same is repeated for all the planes to reconstruct 3D volume of the specimen [23]. Other techniques that are capable of 3D imaging are biplane fluorescence PhotoActivation Localization Microscopy (biplane-fPALM) and double-helix PSF [24, 25]. Few techniques use astigmatism to realize 3D imaging. This includes 3D Stochastic Optical Reconstruction Microscopy (STORM) that uses optical astigmatism to determine both lateral and axial positions of individual molecules [26]. The technique uses a cylindrical lens in the detection path to generate a distinct distribution of point emitters (single molecules) based on their axial position. Molecules in focus exhibit a near Gaussian distribution, whereas molecules at the off-focal plane appear elongated, thereby revealing their location with respect to the molecule at the focal plane. Another similar technique is the integration of DNA-PAINT with astigmatism for 3D imaging [27]. The technique demonstrates a multifold increase in imaging speed with the capability of three-dimensional super-resolution imaging and does not need optical sectioning. A more recent technique is $scanSMLM$, which employs an electrically tunable lens in the detection sub-system, thereby axially scanning the entire specimen. Localizing all the detected molecules gives rise to a 3D map of single molecules. The technique is demonstrated for 10 planes, however, it can be carried out for much larger specimen planes, provided sampling along the axial direction satisfies the sampling theorem and respects the diffraction limit [28].

Photobleaching and scattering are the main hindrances to super-resolve features at large penetration depths. In this respect, SMLM techniques such as fPALM/PALM/STORM demonstrated resolution better than 10–20 nm [29]. However, STED could reach a depth of 20 fold (i.e., up to $10 - 15$ μm) and with a resolution of about 70 nm. This is good, considering the fact that most of the cellular structures can be resolved at much larger penetration depths. In another development, STED microscopy with a hollow Bessel beam for depletion has shown super-resolution at a depth of >100 μm. This is incredible, extending super-resolution at vast depths and will have implications in brain imaging [30].

One of the important and less explored aspects is temporal resolution in SMLM [31]. Although the temporal resolution is decided by the fluorophore's excitation-emission cycle [32], so techniques that can go fast and help reduce photobleaching are of great potential. In general, the spatial and temporal resolution depends on several factors, such as laser intensity, camera speed, and ultimately, the density of molecules detected and discerned in the reconstructed super-resolved image [33]. Though SMLM significantly improves spatial resolution as compared to other super-resolution techniques, it comes with a trade-off in temporal resolution. Recently, single-molecule switching nanoscopy at high speeds has been shown to achieve similar image quality with a 5–25 fold shorter time [34]. In another development, Basumatary et al. demonstrated $shortSMLM$ technique that showed an appreciable improvement in temporal resolution [35]. The technique demonstrated an improvement of 2.76 in temporal resolution; however, it comes with a sacrifice in spatial resolution, giving an overall PArticle Resolution shift (PAR-shift) of 11.82 nm. In general, techniques such as, STED and SIM offer higher temporal resolution (in the order of milliseconds), while SMLM sometimes take longer acquisition times [36]. Over the years, this has improved substantially, but more needs to be done to study dynamic events in a cell.

The ability to resolve events in time is equally important as that of localizing single molecules in space. Specifically, in SMLM, capturing fast-blinking molecules may help reduce the data collection time. When integrated with fast computing engines, this may render real-time imaging, improving temporal resolution. In this respect, fast computing engines (FPGA-GPU and CUDA) coupled with artificial intelligence have shown great potential for improving temporal resolution in SMLM [37–40]. Another approach is to increase the number of active single molecules per frame without changing the intensity as this may increase photobleaching [41,42]. So, rapid data acquisition in SMLM and fast computing engines holds the key for real-time super-resolution imaging.

1.3 The Advent of Super-Resolution in Biological Sciences

Once the optical resolution limit is circumvented, the path to diversification and applicability is opened. This is true for biological science that has embraced and hugely benefited from super-resolution microscopy. Although further resolution improvements are welcome, its applicability and reliability to specific fields take the central stage.

Aspects that elevate the applicability of SMLM is the development of correlative microscopy, which seeks to combine the strengths of complementary imaging techniques. The nanoscale organization of proteins and other components in a cell and their relationship with hundreds of other proteins provides a comprehensive view across the cell volume and helps understand their function. In this respect, the combination of SMLM and cryoelectron microscopy is great for studying cells [43]. The correlative technique allows these details by direct visualization of relationships of specific proteins. In addition, imaging events in crowded cellular environments requires both fluorescence-based information and cryo-EM data. R. Kaufmann's group demonstrated the feasibility of fluorescence imaging (resolution ~400 nm) and its correlation with cryo-EM (resolution in Angstrom range), that has a large resolution gap. This has enabled the direct correlation of structural details of both modalities, allowing deeper investigation of biological processes. In another development, cryo-STED is also attempted [44]. The combination of SMLM and cryoEM is a great combination that is in the making. However, the combination is also carried out by Chang et al. [45]. Still, a number of technical challenges and the lack of knowledge of photo-switching under cryo-conditions are hindering progress.

Super-resolution microscopy is often used to visualize biological structures such as, actin filaments, mitochondria, nuclear pores, microtubules or Golgi bodies and understand disease biology [46]. The fact that fluorescent proteins allow cloning/conjugation with protein-of-interest makes it suitable for studying biological processes and helps in understanding the underlying mechanism. For example, fundamental understanding of nucleosome assisted assembly of higher-order structure and how it affects genome organization is studied using SMLM [47]. In a separate study, super-resolution microscopy (STED and 3D-SIM) techniques are used to understand the role of nanodomain formation in DNA repair and spreading of Xist RNA molecules during X-chromosome inactivation [48, 49].

SMLM has emerged as a key investigation technique for disease biology. For example, Samuel Hess group showed that SMLM can be used to understand the mechanism of action of one of key conserved protein (Hemagglutinin) in Influenza (type-A) infection. In addition, the group suggested an ammonia based compound cetylpyridinium chloride (CPC) showing its antibacterial and antiviral properties and capablility to disrupt the clustering process [50]. In another develpoment, Partha Mondal group used scanning SMLM volume super-resolution microscopy to study NS2B3 and NS2B protein in Dengue (type 2) [51, 52]. Infact, SMLM is also used for drug treatment studies and drug discovery [53–55]. These advances emphasizes the importance of super-resolution microscopy in disease biology.

Lately, super-resolution techniques have penetrated clinical disciplines, including diagnostics. In ophthalmology, structured illumination ophthalmoscope is developed, which combines SIM with an ophthalmoscope to super-resolve human retina [56]. In another application, adaptive optics-assisted SMLM is used in food research. Protein phosvitin (present in egg yolk) is labeled with fluorescent antibodies to obtain super-resolved images of phosvitin-covered oil droplets with 100 nm resolution [57]. This is indeed an excellent example of how super-resolution is shaping fields outside biological sciences.

A timeline depicting the development of the super-resolution concept and that of applications assisted by single molecule based super-resolution microscopy is shown in Fig. 1.1.

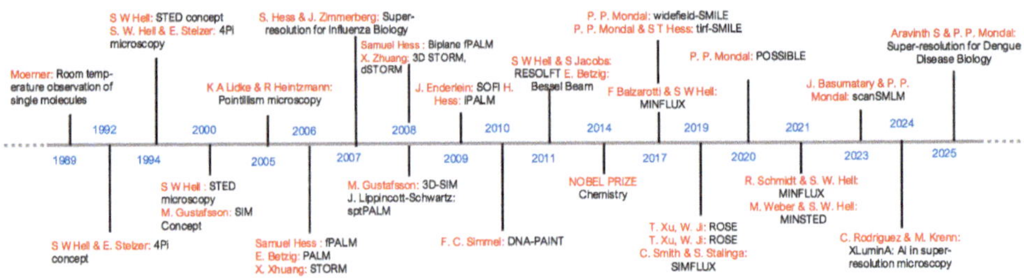

Fig. 1.1 Timelines highlighting critical milestones in the development of single molecule super-resolution microscopy

1.4 What Lies Beyond in Super-Resolution Microscopy

Very recent advances in SMLM indicate the need for high spatio-temporal super-resolution. Such a system can possibly decode biological processes in action without the hangover of limitations in temporal and spatial resolution.

Looking into the future, open technology for super-resolution microscopy is destined to become a reality. This is largely driven by the research community and promoted by commercial interest. In this respect, projects such as, Fiji image analysis software, mobile phone-based SMLM (cellSTORM), and Openflexure microscope have shown the potential of these open source products [58–60]. Open technologies help the scientific community to harvest the potential of SMLM in different fields, thereby diversifying the technique. In addition, an open source "frugal science" is enabling the availability of scientific instruments at minimal costs [61]. The way it is done is by restructuring laboratory-grade devices with a do-it-yourself approach. The open-source documentation helps non-specialists to build high-end super-resolution imaging devices themselves. The cost of the actual commercial SMLM system can be slashed down to about 1/15th for a do-it-yourself approach. Such an arrangement can help researchers in financially-poor institutes to teach and use it for their research.

It is indeed a pity that single molecule techniques have yet to be explored in fields other than biological sciences. The future will likely see a surge of these techniques in physics, chemistry, engineering and medical disciplines as well. It is a testament that new techniques and capabilities have always brought new insights into a field and accelerated its growth. What makes single molecule techniques so unique is the fact that "One can't do better than the smallest functional unit of life i.e, the single molecule"

References

1. Abbe E (1873) Beiträge zur Theorie des Mikroskops und der mikroskopischen Wahrnehmung. Archiv für mikroskopische Anatomie 9:413–468
2. Abbe E (1882) The relation of aperture and power in the microscope (continued). J R Microsc Soc 2(4):460–473
3. Rayleigh LFRS (1879) Investigations in optics, with special reference to the spectroscope. In: The London, Edinburgh, and Dublin Philosophical Magazine and Journal of Science, 5th Series 8, pp 261–274
4. Toraldo di Francia G (1952) Super-gain antennas and optical resolving power. Nuovo Cim 9:426–435
5. Bailey B, Farkas DL, Taylor DL, Lanni F (1993) Enhancement of axial resolution in fluorescence microscopy by standing-wave excitation. Nature 366:44–48
6. Hell S, Stelzer EH (1992) Properties of a 4Pi confocal fluorescence microscope. JOSA A 9:2159–2166
7. Hell SW, Lindek S, Cremer C, Stelzer EH (1994) Measurement of the 4Pi-confocal point spread function proves 75 nm axial resolution. Appl Phys Lett 64:1335–1337

8. Gustafsson MG, Agard DA, Sedat JW (1995) Sevenfold improvement of axial resolution in 3D wide-field microscopy using two objective-lenses. In Wilson T, Cogswell CJ (eds) Three-dimensional microscopy: image acquisition and processing II, vol. 2412. SPIE, Bellingham, WA, pp 147–156
9. Hell SW, Wichmann J (1994) Breaking the diffraction resolution limit by stimulated emission: stimulated-emission-depletion fluorescence microscopy. Opt Lett 19:780–782
10. Gustafsson MG (2000) Surpassing the lateral resolution limit by a factor of two using structured illumination microscopy. J Microsc 198:82–87
11. Heintzmann R, Jovin TM, Cremer C (2002) Saturated patterned excitation microscopy—a concept for optical resolution improvement. JOSA A 19:1599–1609
12. Gustafsson MG (2005) Nonlinear structured-illumination microscopy: wide-field fluorescence imaging with theoretically unlimited resolution. Proc Natl Acad Sci USA 102:13081–13086
13. Betzig E (1995) Proposed method for molecular optical imaging. Opt Lett 20:237–239
14. Dickson RM, Cubitt AB, Tsien RY, Moerner WE (1997) On/off blinking and switching behaviour of single molecules of green fluorescent protein. Nature 388:355–358
15. Hess ST, Girirajan TP, Mason MD (2006) Ultra-high resolution imaging by fluorescence photoactivation localization microscopy. Biophys J 91:4258–4272
16. Betzig E, Patterson GH, Sougrat R, Lindwasser OW, Olenych S, Bonifacino JS, Davidson MW, Lippincott-Schwartz J, Hess HF (2006) Imaging intracellular fluorescent proteins at nanometer resolution. Science 313:1642–1645
17. Rust MJ, Bates M, Zhuang X (2006) Stochastic optical reconstruction microscopy (storm) provides sub-diffraction-limit image resolution. Nat Methods 3:793–796
18. Juette MF, Gould TJ, Lessard MD, Mlodzianoski MJ, Nagpure BS, Bennett BT, Hess ST, Bewersdorf J (2008) Three-dimensional sub–100 nm resolution fluorescence microscopy of thick samples. Nat Methods 5:527–529
19. Huang B, Wang W, Bates M, Zhuang X (2008) Three-dimensional super-resolution imaging by stochastic optical reconstruction microscopy, Science 319:810–813
20. Dertinger T, Colyer R, Iyer G, Weiss S, Enderlein J (2009) Fast, background-free, 3D super-resolution optical fluctuation imaging (SOFI). Proc Natl Acad Sci USA 106:22287–22292
21. Pavani SR, Thompson MA, Biteen JS, Lord SJ, Liu N, Twieg RJ, Piestun R, Moerner WE (2009) Three-dimensional, single-molecule fluorescence imaging beyond the diffraction limit by using a double-helix point spread function. Proc Natl Acad Sci USA 106:2995–2999
22. Hajj B, Wisniewski J, El Beheiry M, Chen J, Revyakin A, Wu C, Dahan M (2014) Whole-cell, multicolor superresolution imaging using volumetric multifocus microscopy. Proc Natl Acad Sci USA 111:17480–17485
23. Legant W et al (2016) High-density three-dimensional localization microscopy across large volumes. Nat Methods 13:359–365
24. Juette M et al (2008) Three-dimensional sub-100 nm resolution fluorescence microscopy of thick samples. Nat Methods 5:527–529
25. Carr AR et al (2017) Three-dimensional super-resolution in eukaryotic cells using the double-helix point spread function. Biophys J 112:1444–1454
26. Huang B, Wang W, Bates M, Zhuang X (2008) Three-dimensional super-resolution imaging by stochastic optical reconstruction microscopy. Science 319:810–813
27. Chung KKH et al (2022) Fluorogenic DNA-PAINT for faster, low-background super-resolution imaging. Nat. Methods 19(5):554–559
28. Basumatary J, Baro N, Joshi P, Mondal PP (2023) Scanning single molecule localization microscopy (scanSMLM) for super-resolution volume imaging. Commun Biol 6:1050
29. Lin D, Gagnon LA, Howard MD, Halpern AR, Vaughan JC (2018) Extended-Depth 3D Super-Resolution Imaging Using Probe-Refresh STORM Author links open overlay panel. Biophys J 114:1980–1987

30. Yu W, Ji Z, Dong D, Yang X, Xiao Y, Gong Q et al (2016) Super-resolution deep imaging with hollow bessel beam STED microscopy. Laser Photon Rev 10:147–152
31. Caron J, Fallet C, Tinevez JY, Moisan L, Braitbart LP, Sirat GY, Shorte SL (2014) Conical diffraction illumination opens the way for low phototoxicity super-resolution imaging. Cell Adhes Migr 8:430–439
32. Mondal PP (2014) Temporal resolution in fluorescence imaging. Front Mol Biosci 1:11
33. Manley S, Gillette JM, Patterson GH, Shroff H, Hess HF, Betzig E, Lippincott-Schwartz J (2008) High-density mapping of single-molecule trajectories with photoactivated localization microscopy. Nat Methods 5:155–157
34. Lin Y et al (2015) Quantifying and optimizing single-molecule switching nanoscopy at high speeds. PLoS ONE 10:e0128135
35. Basumatary J, Baro N, Zanacchi FC, Mondal PP (2023) Temporally resolved SMLM (with large PAR shift) enabled visualization of dynamic HA cluster formation and migration in a live cell. Sci Rep 13:12561
36. Galbraith CG, Galbraith JA (2011) Super-resolution microscopy at a glance. J Cell Sci 124:1607–1611
37. Gui D et al (2021) Accelerating multi-emitter localization in super-resolution localization microscopy with FPGA-GPU cooperative computation. Opt Express 29(22):35247–35260
38. Yuan Y et al (2019) A fast single-image super-resolution method implemented with CUDA. J Real-Time Img Proc 16:81–97
39. Munro I et al (2019) Accelerating single molecule localization microscopy through parallel processing on a high-performance computing cluster. J Microsc 273(2):148–160
40. Jones SA et al (2011) Fast, three-dimensional super-resolution imaging of live cells. Nat Methods 8(6):499
41. Diekmann R et al (2020) Optimizing imaging speed and excitation intensity for single-molecule localization microscopy. Nat Meth 17:909–912
42. Holden SJ, Uphoff S, Kapanidis AN (2011) DAOSTORM: an algorithm for high-density super-resolution microscopy. Nat Methods 8(4):279
43. Hoffmann DP et al (2020) Correlative three-dimensional super-resolution and block-face electron microscopy of whole vitreously frozen cells. Science 367:6475
44. Giske A (2007) CryoSTED Microscopy-A New Spectroscopic Approach for Improving the Resolution of STED Microscopy Using Low Temperature. PhD, Ruperto-Carola University of Heidelberg, Heidelberg
45. Chang YW, Chen SY, Tocheva EI, Treuner-Lange A, Löbach S, Søgaard-Andersen L, Jensen GJ (2014) Correlated cryogenic photoactivated localization microscopy and cryo-electron tomography. Nat Methods 11:737–739
46. Schermelleh L, Ferrand A, Huser T, Eggeling C, Sauer M, Biehlmaier O, Drummen GP (2019) Super-resolution microscopy demystified. Nat Cell Biol 21:72–84
47. Bintu B, Mateo LJ, Su JH, Sinnott-Armstrong NA, Parker M, Kinrot S, Yamaya K, Boettiger AN, Zhuang X (2018) Super-resolution chromatin tracing reveals domains and cooperative interactions in single cells. Science 362:eaau1783
48. Ochs F et al (2019) Stabilization of chromatin topology safeguards genome integrity. Nature 574:571–574
49. Rodermund L et al (2021) Time-resolved structured illumination microscopy reveals key principles of Xist RNA spreading. Science 372:eabe7500
50. Raut P et al (2022) Phosphatidylinositol 4,5 Bisphosphate Mediates the Co-Distribution of Influenza A Hemagglutinin and Matrix Protein M1 at the Plasma Membrane. Viruses 14(11):2509

51. Varghese JM, Joshi P, Aravinth S, Mondal PP (2022) Single-Molecule Super-Resolution Imaging Reveals Formation of NS2B3 Protein Clusters on Mitochondrial Network Leading to its Fragmentation during the Onset of Dengue (Denv-2) Viral Infection. BioRxiv https://doi.org/10.1101/2022.12.18.520514
52. Aravinth S, Varghese JM, Pant N, Mondal PP (2025) Super-resolution optical microscopy reveals accumulation of photoactivable dengue protein (Dendra2-NS2B) in the endoplasmic reticulum. Sci Rep 15:10358
53. Raut P, Obeng B, Waters H, Zimmerberg J, Gosse JA, Hess ST (2022) Phosphatidylinositol 4,5-Bisphosphate mediates the co-distribution of influenza a hemagglutinin and matrix protein M1 at the plasma membrane. Viruses 14(11):2509
54. Raut P et al (2022) Cetylpyridinium chloride (CPC) reduces zebrafish mortality from influenza infection: super-resolution microscopy reveals CPC interference with multiple protein interactions with phosphatidylinositol 4,5-bisphosphate in immune function. Toxicol Appl Pharmacol 440:115913
55. Varghese JM, Aravinth S, Pant N, Mondal PP (2023) The Anti-inflammatory Drug Leflunomide Inhibits NS2B3 Cluster Formation During Dengue Viral Infection as Revealed by Single Molecule Imaging, BioRxiv. https://doi.org/10.1101/2023.12.26.573168
56. Schock F, Best G, Celik N, Heintzmann R, Dithmar S, Cremer C (2022) Structured illumination ophthalmoscope: super-resolution microscopy on the living human eye. Phil Trans R Soc A 380:20210151
57. Jabermoradi A, Yang S, Gobes MI, van Duynhoven JPM, Hohlbein J (2022) Enabling single-molecule localization microscopy in turbid food emulsions. Phil Trans R Soc A 380:20200164
58. Collins JT et al (2020) Robotic microscopy for everyone: the OpenFlexure microscope. Biomed Opt Express 11:2447–2460
59. Schindelin J et al (2012) Fiji: an open-source platform for biological-image analysis. Nat Methods 9:676–682
60. Diederich B, Then P, Jügler A, Förster R, Heintzmann R (2019) cellSTORM—Cost-effective super-resolution on a cellphone using dSTORM. PLoS One 14:e0209827
61. Wang H, Lachmann R, Marsikova B, Heintzmann R, Diederich B (2022) UCsim2: 2D structured illumination microscopy using UC2. bioRxiv. https://doi.org/10.1101/2021.01.08.425840

Basics of Ray Optics for Optical Microscopy 2

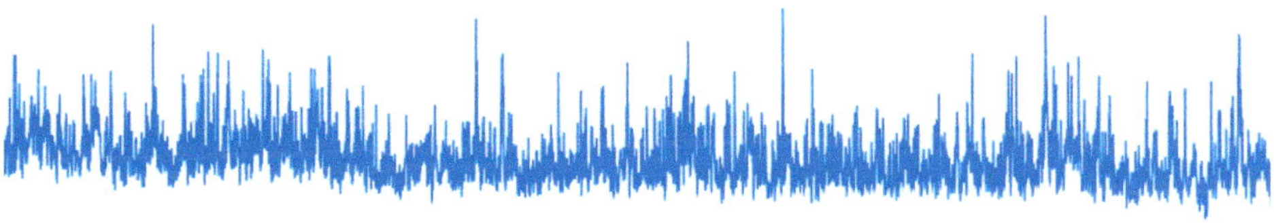

The simplest quantitative understanding of light is facilitated by geometric theory of light also known as Ray/geometric optics. The theory is based on a set of postulates that are found to preserve everyday observations on length scales much longer than the wavelength, and are not violated except under special circumstances. Among the postulates, the most profound being Fermat's principle which governs the propagation of light traveling between two points. Specifically, it states that light takes minimal time to travel between two points, and largely the path of minimum time is also the the path of minimum distance (Hero's law). Fermat's principle can be taken as the governing principle to derive the core laws of ray optics i.e, law of reflection, and law of refraction (Snell's law). This opens the door for predicting the path of light as it propagates through interfaces and key elements of (e.g., mirrors, optical glasses, lenses etc) an optical system. The theory also forms the basis for some of the celebrated simplistic formulas (imaging equation and lens maker's formula). Moreover, this leads to accurate functioning of complex optical imaging components such as, beam expander and objective lens. In addition, we will discuss different aberration effects caused by these optical components. The chapter also discusses some of the the optical components and systems necessary for constructing a high-resolution precision optical microscopy system.

With the development of high-performance optical microscopy systems, geometric optics has taken a central stage. This is largely due to its simplicity and easy interpretation. Although, in principle, a much more complete theory, such as electromagnetic optics

P. P. Mondal, S. Hess, *Single-Molecule Super-Resolution Optical Microscopy*,
https://doi.org/10.1007/978-3-032-13442-4_2

and quantum optics are capable of explaining almost all observations pertaining to light, however they are complex, mathematically hard, and not practical. This is where simplistic theories make the difference and pave the way for finding solutions to everyday observations. Moreover, geometric optics is the most suitable among other theories of light for realizing an optical system. This includes aligning the light beam and understanding its propagation through a diverse range of optical elements. However, complex theories may need to be invoked in the problem formulation and construction of optical systems that involve the wave, electromagnetic and particle nature of light. A rule of thumb could be to use geometric optics when the dimension of objects (such as a lens) through which the light passes is much larger than the wavelength of light, whereas wave theory of light may be more appropriate when the dimension of objects (such as pin-hole) is comparable to the wavelength of light. A simple example would be an imaging system that involves understanding the field strength of an interfering light at the focus of a lens system. Realizing such a system requires a light source emitting a pencil of light beam, which is being directed to the specimen using a series of mirrors, lenses, and grading index glass (glass with varying refractive index). This requires using laws of reflection and refraction to guide the light through lenses and mirrors, whereas Fermat's principle can be directly employed to direct light through gradient index materials. Since the systems involve interference, wave optics can be employed to realize the interference field at the focus of the lens system. However, the electromagnetic theory of light needs to be employed to determine the strength of the electromagnetic field. This clearly shows the reach of each theory, their applicability regime, and combined effort in realizing an optical system.

2.1 Geometric/Ray Optics

The ability of geometric optics to explain everyday observations makes it stand out and a popular choice. The basic assumption is that light behaves like a ray between two points, and simple geometric laws can be employed to understand its behavior (reflection, refraction, and propagation). More specifically, these laws can facilitate prediction of new coordinates (position and direction) of the light rays from the existing coordinate. Practically, this means the determination of beam position and direction as the light bounces off from multiple reflecting surfaces and/or passes through complex lens systems. The theory is primarily based on paraxial approximation, which states that rays passing through optical elements make small angles with respect to the optical axis. This necessitates the rejection of large-angle rays and, as a result, facilitates the simple formulation of optical systems, leading to easy-to-interpret formulas and equations such as lens-makers formula and imaging equation. Within these sets of assumptions and circumstances, a set of postulates governing the behavior of light at macroscopic length scales may be proposed.

Hence, it is always advantageous to begin with the postulates of geometric optics followed by its implications as light passes through optical elements in an imaging systems.

Box 2.1 Postulates of Geometric/Ray Optics

I: Light as Rays. Light travels in the form of rays obeying geometric rules. They emanate from a source and cease at the detector.

II: Optical Path-length. This total distance travelled by light as it pass through a medium or set of mediums is called optical path-length. For the simplistic situation of a single homogeneous medium of refractive index (RI) "n", the time taken to travel a distance "d" is, $t = \frac{d}{c} = \frac{nd}{c_0}$. Here, $c(= c_0/n)$ and c_0 are the speed of light in the medium and vacuum, respectively. The effective path-length (l) is, $l = n \times d$. For a generalized case of inhomogeneous medium (represented by, $n(\vec{r})$) where RI is a function of position, the optical path-length is given by,

$$l = \int_{r_1}^{r_2} n(\vec{r}) dr \tag{2.1}$$

where, r_1 and r_2 are the start and end points of optical path "r", respectively.

III: Fermat's Principle. An optical ray travelling between two arbitrary points (say, r_1 and r_2) follow a path such that the optical path-length or equivalently the travel-time is minimum with respect to other possible paths between r_1 to r_2 i.e,

$$\Delta\left(\int_{r_1}^{r_2} n(\vec{r}) dr\right) = 0. \tag{2.2}$$

where, Δ signifies extremum. In-general, it is the minimum that implies, "*Light rays travel along the path of minimal time or equivalently of shortest optical distance*".

In practice, many imaging elements are manufactured using a single homogeneous medium. In these optical elements, rays follow the path of shortest distance. This is also knows as, **Hero's law**.

The postulates of geometric optics are the guiding principles for determining light beam propagation in and across a medium or multitude of mediums. They dictate the path followed by a light beam as they undergo reflection or refraction or as they propagate through optical elements (such as glass, mirror, lens, and prisms) that are made of different optical materials (glass, polymer, or silica).

2.2 Laws of Reflection and Refraction (Snell's Law)

Our observation of light bouncing off reflecting surfaces, transmission through interfaces, and propagation through multiple mediums led us to believe that it follows certain physical laws. These laws pertain to two different cases: (1) Light as it reflects the surface, and (2) Light passing through interfaces and different mediums. As it turns out, these laws are corollary to Fermat's principle and can be readily derived from the postulates. These laws are significant consequences of Fermat's general theorem and occupy the central stage for the day-to-day interpretation of light propagation. Through examples, we will be able to derive these laws for simpler cases (such as mirrors, air-glass/air-water/air-oil interface), which is sufficient for understanding situations often faced in optical microscopy.

Considering a homogeneous medium, where the path of minimum time is also the path of minimum distance between two points as per Fermat's principle i.e,

$$\Delta\left(\int_{r_1}^{r_2} n(\vec{r})dr\right) = n'\,\Delta\left(\int_{r_1}^{r_2} dr\right). \tag{2.3}$$

where, n' is a constant in the homogeneous medium. Hence the path travelled by light is a straight line connecting r_1 and r_2. This is also illustrated in the example Fig. 2.1, where light travels between two mediums following the laws of reflection and refraction.

In brief, the laws are consequence of Fermat's principle and can be stated as,

Law of Reflection Light travels between two points such that the angle of incidence is equal to the angle of reflection, and both the rays lie in the plane of incidence (see, Fig. 2.1).

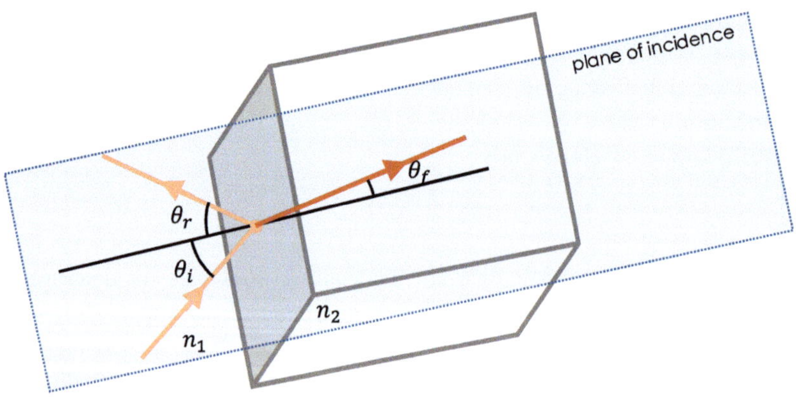

Fig. 2.1 Light propagation between two mediums following Fermat's principle

Using Fermat's principle, one can show that the shortest path between two points (A and B in Fig. 2.1) is the one that has equal incident and reflection angles i.e,

$$\theta_i = \theta_r. \tag{2.4}$$

Law of Refraction (Snell's Law) The transmitted light ray bends and lies in the plane of incidence such that, the following relations holds between two mediums (1 and 2) i.e,

$$n_i \sin\theta_i = n_2 \sin\theta_f \tag{2.5}$$

where, n_1 and n_2 are the refractive indices of two mediums. Again, Fermat's principle ensures that the path between the two points (A and C) is minimum. This is possible only if the travel time is more in lower refractive index (RI) medium and less in high RI medium, ensuring minimum pathlength between two points. The same is shown in Fig. 2.1).

In this chapter and rest of the book, we will take these laws as accepted and build upon it. Interested readers may refer to a rigorous optics book for further details and the derivation of Eqs. 2.4 and 2.5 [1–3].

2.3 Paraxial Ray Approximation

A simplistic formulation of ray optics is possible only if certain approximations are considered in a wide range of diverse conditions faced by light as it propagates. One such approximation is "*Paraxial Ray Approximation*" that applies to light rays near the optical axis. This gives a simplistic interpretation of light as it propagates through different mediums/optical components and facilitates easy problem formulation. Formally, the approximation states that, " *Paraxial rays are the rays that make small angles with the optical axis and as a consequence,* $\sin\theta \approx \theta$ *holds well*". It may be noted that paraxial rays hold good for ideal conditions where rays are near parallel to the optical axis. So, the approximation is valid for rays making small angles with the optical axis that largely takes place in close proximity to the center of optical components such as mirrors, lenses, etc. In practice, paraxial rays are used to determine imaging parameters (image location and magnification) of an optical component or system

2.3.1 Spherical Mirror

One of the simplest examples where the paraxial approximation is extensively used is a spherical mirror, specifically near the optical axis, where it approximates a paraboloidal mirror that focus rays on a single point. Figure 2.2 shows the propagation of rays from an object (OO') to the image (II') after getting reflected by a spherical mirror. Light from the object point O and O' (at a distance v from the mirror) hits the mirror and gets reflected

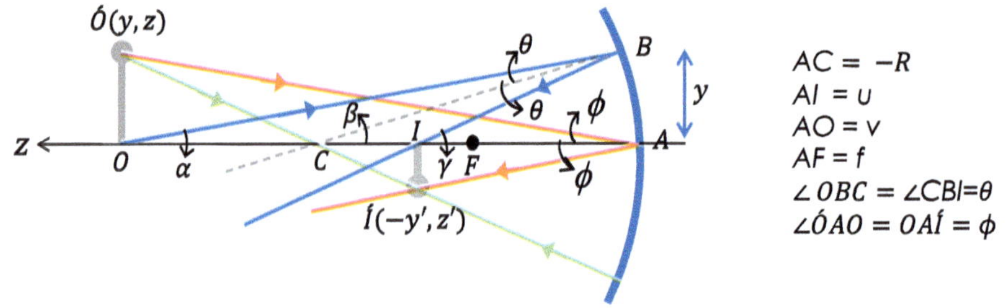

Fig. 2.2 Reflection and image formation by a spherical mirror

as per Fermat's principle (more appropriately, Hero's law) to the image points I and I', respectively which is at a distance u from the mirror. According to the laws of reflection, the angle of incidence (OBC) and reflection (CBI) are the same i.e., $\angle OBC = \angle CBI = \theta$. Similarly, the angle of incidence and reflection at point A are same i.e, $\angle O'AO = \angle OAI' = \phi$ The mirror has a focus of f and a radius-of-curvature R as illustrated in the Fig. 2.2.

For the sake of simplicity, we choose to work with rays that make small angles with the optical axis i.e., paraxial rays. This allows us to write, $\sin\theta \approx \theta$ and $\cos\theta \approx 1$. In the case of spherical mirror, we seek a relation between the object distance, v and image distance, u following general sign convention. To do so, we note the following relations (see, Fig. 2.2),

$$\begin{cases} \beta = \alpha + \theta \\ -\gamma = \beta + \theta \end{cases} \tag{2.6}$$

Combining the above equation we get,

$$-\gamma + \alpha = 2\beta \tag{2.7}$$

Considering paraxial approximation, one can reduce the following trigonometric relations to,

$$\begin{cases} \tan\alpha \approx \alpha = y/v \\ \tan\beta \approx \beta = y/(-R) \\ \tan(-\gamma) \approx -\gamma = y/u \end{cases} \tag{2.8}$$

Now, using the above relations (Eq. 2.8) in Eq. 2.7 produces,

$$\frac{1}{u} + \frac{1}{v} = \frac{1}{(-R/2)}$$

$$\frac{1}{u} + \frac{1}{v} = \frac{1}{f} \tag{2.9}$$

where, $f = -R/2$ is the focal length of the mirror.

This is the celebrated imaging equation for a spherical mirror. The above equation simply states that the image position of an object (located at a distance v) can be determined exactly (provided f is known).

Another aspect is the magnification/demagnification of the object produced by the spherical lens. Using simple geometric relation and applying it on $\Delta OO'A$ and $\Delta II'A$ in Fig. 2.2 one can deduce that,

$$\frac{y}{z} = \frac{-y'}{z'} \Rightarrow m = \frac{y'}{y} = -\frac{z'}{z} \tag{2.10}$$

where, the negative sign indicates inverted image, and $|m|$ as the magnification factor.

In the case of planar mirror (i.e, $R \to \infty$), $u = -v$. This suggests a symmetric relation between the image and object distances, hence they are also called conjugate planes.

Note that, the same relation can be cohesively represented in matrix form as,

$$\begin{bmatrix} y' \\ \theta' \end{bmatrix} = \begin{bmatrix} m & 0 \\ (1/f) & 1 \end{bmatrix} \begin{bmatrix} y \\ \theta \end{bmatrix} = M \begin{bmatrix} y \\ \theta \end{bmatrix} \tag{2.11}$$

where, $\theta = \alpha$ is the input angle, and $\theta' = \gamma$ is the output angle. In matrix formulation, the input and output ray coordinates are represented by, $\begin{bmatrix} y \\ \theta \end{bmatrix}$ and $\begin{bmatrix} y' \\ \theta' \end{bmatrix}$, respectively, whereas M represents the optical element (here, concave spherical mirror).

2.3.2 Biconvex Thin Lens

Another extensively used optical component in microscopy is lens (both biconvex and biconcave). These components decide how tight one can focus the light on the sample. A lens can be thought of as the merger of two spherical interfaces, as shown in Fig. 2.3.

We begin with a simple spherical interface as shown in Fig. 2.3. A light ray from a point A (on the optical axis) strikes the interface at a point and gets refracted as it enters from low (n) to high (n') refractive index medium. Using Snell's law (see, Fig. 2.3 (top left)), we have the following relation between the incident and refracted rays,

$$n \sin(\theta + \beta) = n' \sin(\beta + \theta') \tag{2.12}$$

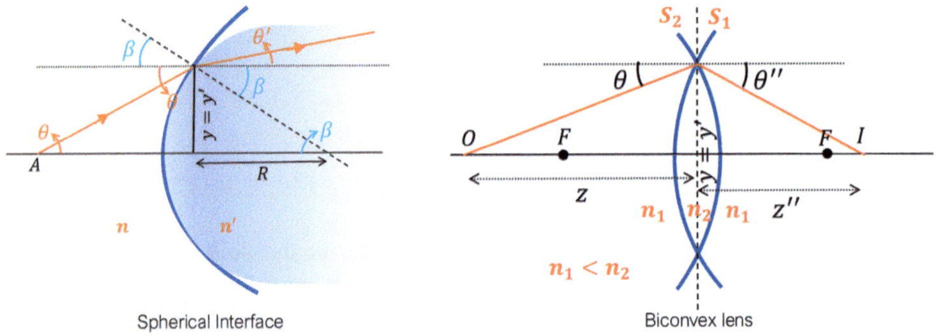

Fig. 2.3 Refraction at a spherical surface. Image formation by a spherical lens

Using Paraxial ray approximation, we get,

$$n\,(\theta + \beta) = n'\,(\beta + \theta')$$

$$\theta' = \frac{n}{n'}\,\theta - \frac{n' - n}{n'}\,\beta \tag{2.13}$$

Noting that, $\beta \approx y/R$ and using $\Delta n = n' - n$, in the above equation produces,

$$\theta' = \frac{n}{n'}\,\theta - \frac{\Delta n}{n'R}\,y \tag{2.14}$$

Incorporating $y' = y$ and encapsulating the above equation in matrix-form, we have,

$$\begin{bmatrix} y' \\ \theta' \end{bmatrix} = \begin{bmatrix} 1 & 0 \\ -\Delta n/n'R & n/n' \end{bmatrix} \begin{bmatrix} y \\ \theta \end{bmatrix} = M \begin{bmatrix} y \\ \theta \end{bmatrix} \tag{2.15}$$

The above equation (Eq. 2.15) relates the input ray to the output ray coordinates for a spherical interface.

A biconvex lens can be considered as a combination of two spherical interfaces (S_1 and S_2) as shown in Fig. 2.3. A light ray travelling from point O to I need to cross two interfaces of refractive indices, $n_1 \rightarrow n_2$ and $n_2 \rightarrow n_1$. Analyzing the ray from first to second spherical interface i.e, across S_1 produces,

$$\begin{bmatrix} y' \\ \theta' \end{bmatrix} = \begin{bmatrix} 1 & 0 \\ -\Delta n/n_2R_1 & n_1/n_2 \end{bmatrix} \begin{bmatrix} y \\ \theta \end{bmatrix} = M_1 \begin{bmatrix} y \\ \theta \end{bmatrix} \tag{2.16}$$

where, $\Delta n = n_2 - n_1$, with $n = n_1$ and $n' = n_2$.

In a similar manner, the rays across the interface S_2 (from n_2 to n_1) gives,

$$\begin{bmatrix} y' \\ \theta' \end{bmatrix} = \begin{bmatrix} 1 & 0 \\ \Delta n/n_1 R_2 & n_2/n_1 \end{bmatrix} \begin{bmatrix} y \\ \theta \end{bmatrix} = M_2 \begin{bmatrix} y \\ \theta \end{bmatrix} \tag{2.17}$$

where, we have interchanged n and n', and replaced the radius with $-R_2$ for the second surface S_2.

Considering thin lens, we neglect the thickness of the lens with respect to the image and object distances. The matrix for the biconvex lens is simply the multiplication of both the matrices in the reverse order (as per Ray-Matrix formulation) i.e.,

$$M = M_2 \times M_1 = \begin{bmatrix} 1 & 0 \\ \Delta n/n_1 R_2 & n_2/n_1 \end{bmatrix} \times \begin{bmatrix} 1 & 0 \\ -\Delta n/n_2 R_1 & n_1/n_2 \end{bmatrix}$$

Hence,

$$\begin{bmatrix} y' \\ \theta' \end{bmatrix} = \begin{bmatrix} 1 & 0 \\ -1/f & 1 \end{bmatrix} \begin{bmatrix} y \\ \theta \end{bmatrix} \tag{2.18}$$

where,

$$\frac{1}{f} = \frac{\Delta n}{n_1} \left(\frac{1}{R_1} - \frac{1}{R_2} \right) \tag{2.19}$$

Note that, Eq. 2.18 is the imaging equation for a biconvex lens, and more importantly, Eq. 2.19 is the popular **lens-maker's formula**.

2.4 Combination of Optical Components

Beam expansion and focusing are key aspect of optical microscopy and imaging systems. This is achieved by the combination of two or more optical elements. Some of the well-known examples are beam-expander and objective lens. In these conjugate system, matrix formulation comes handy. We will explore some of these combinatorial systems in this section.

2.4.1 Beam-Expander

As the name suggests, beam-expander is an optical system that can expand the width of a collimated light beam to a desired beam waist. A typical beam-expander system consists

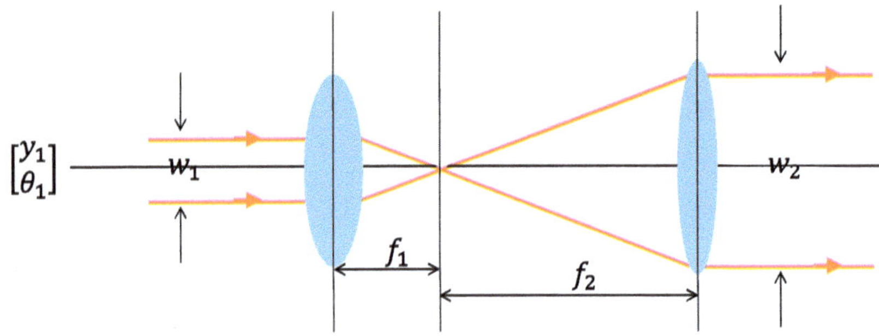

Fig. 2.4 Beam-expander comprising of two biconvex lenses

of two biconvex lenses as shown in Fig. 2.4. However, other combinations exists involving both biconcave and biconvex lenses.

Consider a collimated beam of beam-waist w_1 as input as shown in Fig. 2.4. The input beam of waist w_1 is represented by a ray-vector, $\begin{bmatrix} y_1 \\ \theta_1 \end{bmatrix}$, where, $y_1 = w_1/2$. Let us represent the output beam-waist by, $\begin{bmatrix} y_2 \\ \theta_2 \end{bmatrix}$, where, $y_2 = w_2/2$. As per matrix-formulation, the beam-expander system can be decomposed into three sub-systems: the first lens (of focal length f_1), the intermediate space (of distance D), and the second lens (of focal length f_2). Note that, the intermediate space D is the distance between two lenses which are placed at a distance $f_1 + f_2$.

Representing each sub-system by a matrix i.e, first lens by M_1, intermediate space by T, and second lens by M_2, the matrix for the composite system is,

$$M = M_2 \times T \times M_1 \tag{2.20}$$

where, the matrices are given by,

$$M_1 = \begin{bmatrix} 1 & 0 \\ -1/f_1 & 1 \end{bmatrix}$$

$$M_2 = \begin{bmatrix} 1 & 0 \\ -1/f_2 & 1 \end{bmatrix}$$

$$T = \begin{bmatrix} 1 & D \\ 0 & 1 \end{bmatrix}$$

Incorporating the individual matrices gives the following matrix for the whole system,

$$M = \begin{bmatrix} \left(1 - \frac{D}{f_1}\right) & D \\ \frac{1}{f_1} + \frac{1}{f_2}\left(1 - \frac{D}{f_1}\right) & \left(1 - \frac{D}{f_2}\right) \end{bmatrix} \tag{2.21}$$

So, the relation between input and output ray-vectors become,

$$\begin{bmatrix} y_2 \\ \theta_2 \end{bmatrix} = M \times \begin{bmatrix} y_1 \\ \theta_1 \end{bmatrix} \tag{2.22}$$

This is a general expression for a beam-expander system. Note that, $D = f_1 + f_2$, and using the same the relation between input and output ray elements are given by,

$$\begin{cases} y_2 = (1 - D/f_1)y_1 + D\theta_1 \\ \theta_2 = -\frac{f_2}{f_1}\,\theta_1 \end{cases} \tag{2.23}$$

In this case, $\theta_1 = \theta_2 = 0$. So, we have the following simplified form,

$$\begin{cases} y_2 = (1 - D/f_1)y_1 \\ \theta_2 = \theta_1 = 0 \end{cases}$$

Again using, $D = f_1 + f_2$, we get the following expression for magnification,

$$\frac{y_2}{y_1} = -\frac{f_2}{f_1}$$

Noting that, $y_1 = w_1/2$ and $y_2 = w_2/2$, the above relation becomes,

$$\frac{w_2/2}{w_1/2} = -\frac{f_2}{f_1}$$

This gives the magnification or expansion of the beam by a factor,

$$m = \left|\frac{w_2/2}{w_1/2}\right| = \left|\frac{f_2}{f_1}\right| \tag{2.24}$$

So, the output beam waist is,

$$w_2 = m\ w_1 \tag{2.25}$$

The above relation is the key relation for a beam-expander system, where the output beam is expanded by a factor $m = |f_2/f_1|$. Beam-expander system is often used in an optical microscopy system to broaden the beam so that it fills the back-aperture of objective lens for utilizing maximum aperture angle and tight focusing.

2.4.2 Objective Lens

Another key component of an optical microscope is the focusing element which is the objective lens. These optical elements can have very large magnification ($\sim 200X$) that are essential for observing tiny objects such as, cell organelles and structures. Such high magnification can only be achieved by combining several optical elements. So, objective lenses are essentially a complex arrangement of lenses (of different kinds) and half-balls.

Among others, Amici objectives are among the high-end objective lenses, and simplest to understand. Figure 2.5 shows a simplistic example of a typical Amici objective. The first two lens combination of biconvex-planoconcave (BC-PC) is used to gradually bending the beam in a stepwise manner. Finally, a half ball is used to sharply bend the light rays to achieve an overall large aperture angle and generate a tight focus.

The ray matrices for both the BC-PC combination is given by,

$$M_{BC-PC} = M_{PC} \times M_{BC} = \begin{bmatrix} 1 & 0 \\ -1/f_{PC} & 1 \end{bmatrix} \times \begin{bmatrix} 1 & 0 \\ -1/f_{BC} & 1 \end{bmatrix} \tag{2.26}$$

Note that, half-ball can be considered as a convex lens with first surface (convex) of positive radius and the second surface (plane) of infinite radius of curvature i.e, $\frac{1}{f_{HB}} = \lim_{R_2 \to \infty} \frac{\Delta n}{n}(\frac{1}{+R_1} - \frac{1}{R_2}) = \frac{\Delta n}{nR_1}$ (using lens-maker's formula, Eq. 2.19). Incorporating all

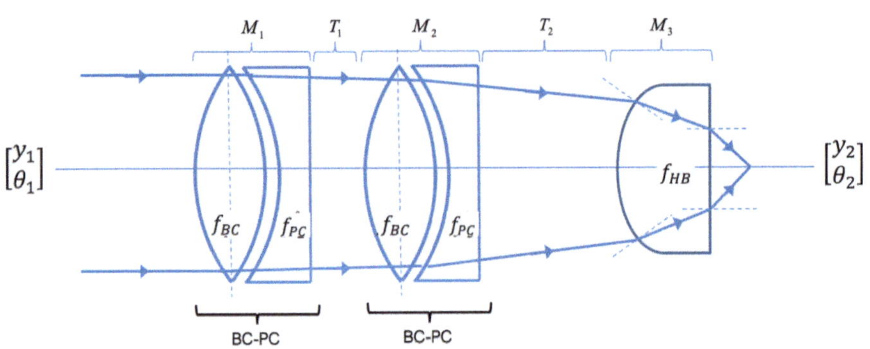

Fig. 2.5 Amici Objective lens system that is composed of two biconvex, two planoconcave and half-ball

the matrix elements, the ray matrix for the half-ball is,

$$M_{HB} = \begin{bmatrix} 1 & 0 \\ -\frac{1}{f_{HB}} & 1 \end{bmatrix} = \begin{bmatrix} 1 & 0 \\ -\Delta n/nR_1 & 1 \end{bmatrix} \tag{2.27}$$

The empty spaces between the optical elements (i.e., between two MC-PC and between BC-PC & HB) is marked in Fig. 2.5 by T_1 and T_2. The corresponding matrices are,

$$T_1 = \begin{bmatrix} 1 & d_1 \\ 0 & 1 \end{bmatrix} \tag{2.28}$$

and,

$$T_2 = \begin{bmatrix} 1 & d_2 \\ 0 & 1 \end{bmatrix} \tag{2.29}$$

Incorporating all the matrices, the matrix for an Amici objective is given by,

$$\begin{aligned} M_{Obj} &= M_{HB} \times T_2 \times M_{BC-PC-2} \times T_1 \times M_{BC\ PC-1} \\ &= \begin{bmatrix} 1 & 0 \\ -1/f_{HB} & 1 \end{bmatrix} \times \begin{bmatrix} 1 & d_2 \\ 0 & 1 \end{bmatrix} \times \begin{bmatrix} 1 & 0 \\ (\frac{1}{f_{PC2}} - \frac{1}{f_{BC2}}) & 1 \end{bmatrix} \\ &\times \begin{bmatrix} 1 & d_1 \\ 0 & 1 \end{bmatrix} \times \begin{bmatrix} 1 & 0 \\ (\frac{1}{f_{PC1}} - \frac{1}{f_{BC1}}) & 1 \end{bmatrix} \end{aligned} \tag{2.30}$$

This is the ray-matrix of a simple objective lens with a moderate magnification. In practice, objectives are highly complex optical elements with several optical components. These lenses are designed to serve specific purpose in optical microscopy such as, high magnification, large working distance, large depth-of-focus etc.

2.5 Optical Aberrations

Finally, we briefly discuss aberration in optical systems which originates mostly due to the lens aperture, symmetry around the axis and other factors. Rather than deriving the grand equation, we will take it and discuss different types of aberrations encountered in optical microscopy in general and super-resolution microscopy in particular. Complete derivation of the grand equation can be found in Ref. [4].

The grand equation for describing aberration in a lens system is given by [4],

$$\zeta = c_1\, d^4 + c_2\, h^2 d^2 \cos^2\gamma + c_3\, h d^3 \cos\gamma + c_4\, h^3 d \cos\gamma + c_5\, h^2 d^2 \qquad (2.31)$$

where, d characterizes the aperture of the lens, γ as the angle around the axis, and h is the distance from the optical axis.

Here, the *first term* denotes spherical aberration which is the failure of lens to bring all light rays to the same focus uniformly. This is strongly related to the aperture of the lens as apparent from d^4 dependence. The second type of aberration (denoted by *second term*) is astigmatism where rays propagating in two different perpendicular axes have different foci. Clearly, astigmatism strongly depends on the aperture (denoted by d^2 factor) and distance (h) from the optical axis, apart from its dependence on symmetry (γ) around the axis. The third kind of aberration is *coma*, represented by *third term*. This is caused by rays from off-axis, and strongly depends on the aperture d, creating a comet-like blur. The *fourth term* represents *distortion* that has strong dependence on the distance from the optical axis. This is largely due to its deviation from rectilinear projection. Finally, the *figth term* represents *field curvature* which is an aberration caused by spread of focus away from optical axis. So, the object cannot be focused on a flat imaging sensor. All these aberrations are displayed in Fig. 2.6.

With the brief introduction to the essential concepts and the behaviour of light passing through various optical elements, we have the essential ray optics background necessary to build a super-resolution microscope. In the next chapter, we will focus on the wave aspects of light that are essential for harvesting wave characteristics of light in high-resolution optical microscopy.

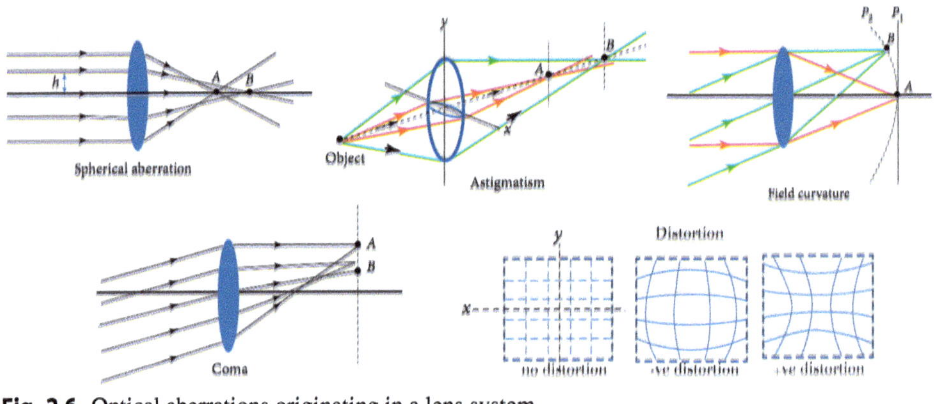

Fig. 2.6 Optical aberrations originating in a lens system

References

1. Saleh BEA, Teich MC (2012) Fundamentals of photonics, 2nd edn. Wiley, New York
2. Born M, Wolf E (1999) Principles of optics, 7th edn. Cambridge University, Cambridge
3. Mondal PP (2021) Classical and quantum optics. American Institute of Physics Publisher, New York
4. Mondal PP (2022) Chapter 1, Book: classical and quantum optics, 1st edn. American Institute of Physics Publishers, New York

Diffraction of Light and Classical Resolution Limit 3

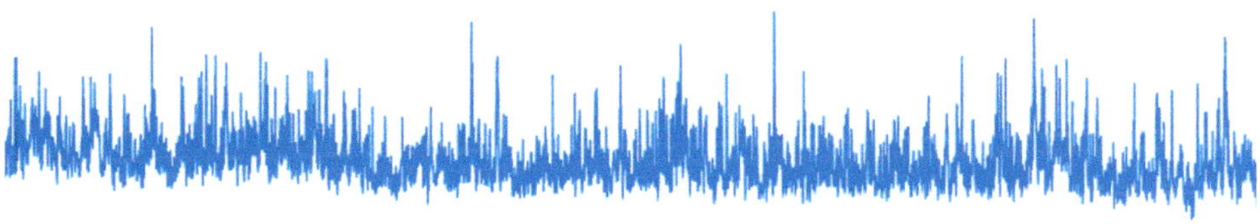

One of the most profound properties of light is its wave nature and the ability to diffract. This refers to the deviation of light from a straight line during propagation and is apparent when light passes through sharp edges. This immediately makes us wonder whether light is more than a geometric ray and prompts us to consider that it is also a wave just like water or sound waves. So, the simplest yet complete way to understand the diffraction nature of light and its properties is to consider it as a wave. Accordingly, we discuss the wave theory of light based on a set of postulates comprising the wave equation and light wave propagation in different mediums. Subsequently, the wave theory is used to understand the diffraction of light. Specifically, we will focus our discussion on the diffraction of light by an objective lens, which is central to super-resolution microscopy. In addition, the diffraction property will help us determine the minimum solvable distance i.e., resolution limit (Abbe/Rayleigh limit). This will identify the factors affecting the resolution limit of an optical microscope. The chapter will set the stage for numerous microscopy systems capable of super-resolution.

The fact that light diffracts and interferes leads us to believe that light also behaves like a wave apart from the fact that it behaves like a geometric ray. Diffraction and interference of light are readily observed in simple experiments. For example, consider light passing through a small hole with an observation screen placed at a distance. One can observe that the pattern on the screen is a magnified version of the pinhole, along with concentric circles of diminishing intensity. In other instances, we can observe the superposition of two nearby

P. P. Mondal, S. Hess, *Single-Molecule Super-Resolution Optical Microscopy*,
https://doi.org/10.1007/978-3-032-13442-4_3

cylindrical waves generated by dropping stones in a pond. At a later time point, the mixing of waves gives rise to elevated waves (crests) or diminishing waves (troughs). This goes on to say that the superposition of two waves can be more than the sum of both. The process is generally termed as interference with crests and troughs resulting as a consequence of constructive interference, whereas the nodes result from destructive interference of waves.

In order to understand the diffraction of light, we need a theory that describes how light propagates near and far from the source. Specifically, our focus will be on the diffraction of light at the far-field, i.e., when the distance is much larger than the wavelength of light, $d >> \lambda$. In fact, it is the diffraction of light at the far-field (caused by the optical focusing element) that determines how fine the features (in the sample) can be resolved. A simple system would be an optical system where diffraction from a circular aperture is focussed by an optical element (objective lens). These kind of systems are often used in optical microscopy to determine the resolving power. Obviously, resolvability depends on the wavelength of light and focusability of the objective lens (numerical aperture). All these aspects will be the subject matter of this chapter.

3.1 Postulates of Wave Optics

The wave theory of light is strongly grounded on the following founding postulates:

I. **Wave Nature:** Light behaves wave (just like water and sound waves), and propagates in both space and time.

II. **Wave Equation:** An optical wave (light) can be conveniently represented by a scalar wavefunction,[1] $\psi(\vec{r}, t)$ that satisfies the wave equation i.e,

$$\left[\frac{1}{c^2}\frac{\partial^2}{\partial t^2} - \nabla^2\right]\psi(\vec{r}, t) = \Box^2\,\psi(\vec{r}, t) = 0 \tag{3.1}$$

where, $\Box^2$ and ∇^2 are the d'Alembert and Laplacian operators' respectively.

III. **Superposition Principle:** The solutions of the wave equation are possible optical waves (say, $\psi_i(\vec{r}, t)$) and obey the principle of superposition i.e,

$$\psi(\vec{r}, t) = \sum_{i \in S} \psi_i(\vec{r}, t) \tag{3.2}$$

where, S is the solution set.

IV. **Light in a Medium:** An optical wave (light) undergoes reflection and refraction at an interface of mediums with different refractive indices. Light travels with a speed of $c = c_0/n$ in a medium, where n is the refractive index of the medium and $c_0 = 2.99792458 \times 10^8$ m/s is the speed in vacuum.

[1]However, light is truly a vector quantity with $\vec{E}$ and $\vec{B}$ field vectors which will be apparent in the next chapter.

The above postulates are well-founded and time-tested. To date, the postulates are well-preserved. In short, the scalar wavefunction representing optical wave/light has been well-behaved under most conditions other than when polarization is important. The wavefunction, $\psi(\vec{r}, t)$ is a function of space (x, y, z) and time, t. So, any solution that satisfies the wave equation is a possible optical wave and, of course, its superposition sum is a solution by virtue of postulate IV.

3.2 Plane Wave

In analogy with other scalar wavefunctions representing water and sound waves, the solution of the wave equation must be periodic in space and time. This means that the solution is a periodic function of space $(\vec{r})$ at a specific time point and a periodic function of time at a particular location in space. Overall, the wavefunction representing a light wave is a periodic function of both space and time.

For simplicity, we begin with an idealistic plane wave that is both monochromatic and periodic. A simple analogy is with water waves, which, when disturbed, give rise to waves with a well-defined wavelength (crest-crest or trough-trough distance) that appear as equidistance parallel planes (far from the origin of spherical waves). Note that the wave oscillates in time at any specific space-point and propagates in space as the wave spreads out. A simplistic way of explaining this nature is by mathematically modeling the waves by periodic function such as *Sine/Cosine* function that incorporates space and time as its variables i.e,

$$\psi(\vec{r}, t) = \psi_0 \cos(\vec{k}.\vec{r} - \omega t + \phi) \tag{3.3}$$

Next we concentrate on the time-component, and rewrite the above relation as,

$$\psi(\vec{r}, t) = \psi_0 \cos(\omega t + \phi) \tag{3.4}$$

where, ψ_0 and ϕ are respectively the amplitude and phase at the location $\vec{r}$.

To bring out the periodic nature of wavefunction, we rewrite the above equation in the following form,

$$\begin{aligned}\psi(\vec{r}, t) &= \psi_0 \cos[\omega(t + \phi/\omega)] \\ &= \psi_0 \cos\left[2\pi\left(\frac{t}{1/\nu}\right) + \phi\right] \\ &= \psi_0 \cos\left[2\pi\left(\frac{t}{T}\right) + \psi\right]\end{aligned} \tag{3.5}$$

where, $T = 1/\nu = 2\pi/\omega$.

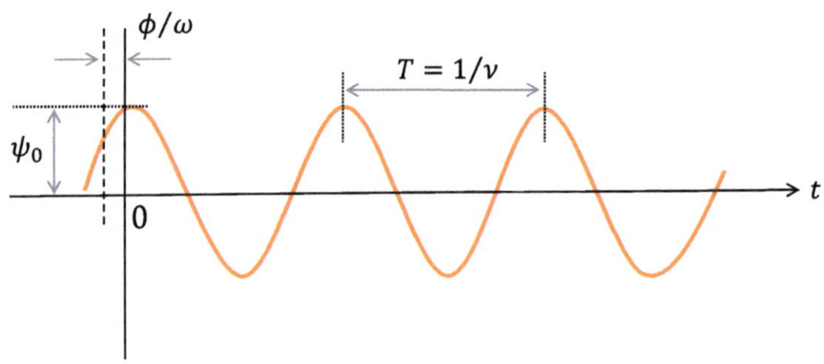

Fig. 3.1 A typical representation of a plane monochromatic wave. All the relevant parameters (time-period, phase etc) of the wave are marked

The optical wave represented by the above wavefunction has an angular frequency of $\omega = 2\pi\nu$, with ν as the frequency (cycles/sec or Hertz). It is evident from the *Cosine* function that the wave repeats itself after every integer multiple of t i.e, $t = T, 2T, 3T, \ldots$. Another critical aspect of the above wave is ϕ, which is a constant offset in the phase of the wave. It is apparent that, $\phi = \phi_0$ at $t = 0$, indicating non-zero phase i.e, the wave is lagging behind with a non-zero lag of ϕ/ω in space as shown in Fig. 3.1.

In general, the wavefunction can be represented in exponential form as,

$$\psi(\vec{r}, t) = \psi(\vec{r})\, re(e^{-i\omega t} e^{-i\phi(\vec{r})}) \tag{3.6}$$

where the real-part re(*) can be retrieved whenever needed.

Note that, both the waveforms (Eqs. 3.5 and 3.6) satisfies the wave-equation (Eq. 3.1) and will be interchangeably used in the rest of the book.

3.3 The Helmholtz Equation and Its Solution

3.3.1 Rectangular Coordinate System

The space and time variables in the wave equation are independent and so they can be separated out without the loss of generality. This facilitates the decoupling of both variables and allows independent analysis. Accordingly, we can rewrite the wavefunction as,

$$\psi(\vec{r}, t) = [\psi_0(\vec{r})\, e^{-i\phi(\vec{r})}]\, e^{-i\omega t} = \psi_0'(\vec{r})\, e^{-i\omega t} \tag{3.7}$$

Substituting the wavefunction (Eq. 3.7) in the wave equation (Eq. 3.1) gives,

$$\nabla^2 \psi_0'(\vec{r}) - \frac{1}{c^2}\frac{\partial^2 \psi_0'(\vec{r})}{\partial t^2} = 0 \tag{3.8}$$

The above equation involves only the space variable and it is popularly known as **Helmholtz equation**.

Next, we proceed to determine the solutions of Helmholtz equation in rectangular coordinate system. The first step is to assume separability of space variables (x, y, z), and write the spatial wavefunction as,

$$\psi_0'(\vec{r}) = \psi_I(x)\ \psi_{II}(y)\ \psi_{III}(z) \tag{3.9}$$

Substituting the above equation (Eq. 3.9) in the Helmholtz equation (Eq. 3.8) produces a set of three independent partial differential equations,

$$\frac{\partial^2}{\partial q^2}\psi_j + k_q^2\ \psi_j^2 = 0 \tag{3.10}$$

where, $(q, j) = (x, I),\ (y, II),\ (z, III)$ represents wave equation for spatially-separated variables, x, y, z. Here, $k^2 = \sum_j k_j^2$.

Using trial method and assuming solution of the form, $\psi_j = C\ e^{mu}$, produces the following solution,

$$\begin{cases} \psi_I(x) = c_1\ e^{-ikx} + c_2\ e^{+ikx} \\ \psi_{II}(y) = c_3\ e^{-iky} + c_4\ e^{+iky} \\ \psi_{III}(z) = c_5\ e^{-ikz} + c_6\ e^{+ikz} \end{cases} \tag{3.11}$$

Taking only the positive direction (say, $\vec{r}$), the solution can be expressed as,

$$\psi^+(\vec{r}) = b_0\ e^{-i(k_x x + k_y y + k_z z)} = b_0\ e^{-i\vec{k}.\vec{r}} \tag{3.12}$$

where, $b_0 = c_1 c_3 c_5$ is the new constant. Here, $\vec{k} = (k_x, k_y, k_z)$, and $\vec{r} = (x, y, z)$ are respectively the wave propagation vector and position vector.

The above solution (Eq. 3.12) is the most general solution of wave equation in rectangular coordinates, representing waves propagating in the positive direction. A similar expression exists for waves propagating in the negative direction i.e,

$$\psi^-(\vec{r}) = b_1\ e^{i\vec{k}.\vec{r}} \tag{3.13}$$

In the next subsection, we will discuss solution of wave equation in a spherical coordinate system.

3.3.2 Interpretation

Next, we proceed to interpret the solution (Eq. 3.12) and determine key parameters. Note that, the above solution has two important parameters: amplitude and phase given by,

$$\begin{cases} Amplitude: \ |b_0| \\ Phase: \ -\vec{k}.\vec{r} + arg(b_0) \end{cases} \tag{3.14}$$

where, $arg(b_0)$ is the phase angle of the complex variable b_0.

Neglecting the complex constant $arg(b_0)$, we note the periodic nature of wave which ensures that the wavefront $(\vec{k}.\vec{r})$ must obey,

$$\vec{k}.\vec{r} = m\ 2\pi \tag{3.15}$$

where, m is an integer. Note that the above condition ensures periodicity of the wavefunction.

The above condition (Eq. 3.15) is due to the periodic nature of wave that repeats itself after every 2π. The equation can be interpreted as parallel planes (represented by, $\vec{r}$) orthogonal to the propagation vector $\vec{k}$, and the consecutive planes are separated by $\lambda = 2\pi/k$ (see, Fig. 3.2). Mathematically, the above Eq. 3.15 represents parallel planes. This can be easily verified by computational means using Matlab.

To understand wave propagation in space and time, let us consider single space variable (say, z) and time variable t. The resultant wave is a traveling-wave in the positive z-direction i.e.,

$$\psi(z) = |b_0|\ e^{-ik_z z} \tag{3.16}$$

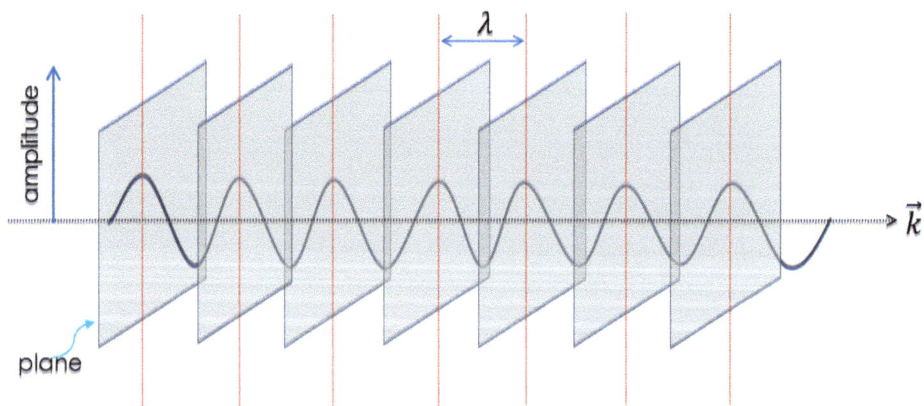

Fig. 3.2 A typical representation of a plane wave showing planes are separated by λ

This gives rise to a complete wavefunction that incorporates time and space variables i.e.,

$$\begin{aligned}\psi(z,t) &= \frac{1}{2}\,[\psi(z)\,e^{i\omega t} + \psi^*(z)\,e^{-i\omega t}] \\ &= \frac{b_0 b_0^*}{2}\,[e^{i(\omega t - k_z z)} + e^{-i(\omega t - k_z z)}] \\ &= |D|\,[\cos(\omega t - k_z z)] \end{aligned} \tag{3.17}$$

where, $|D| = b_0^* b_0$, and the constant phase factor $arg(D)$ is ignored.

The above expression (Eq. 3.17) can be rewritten as,

$$\psi(z,t) = |D|\,\cos\left[2\pi\left(\frac{t}{T} - \frac{z}{\lambda}\right)\right] \tag{3.18}$$

where, $T = 1/\nu$ and $\lambda = c/\nu$ are respectively the time period and space period of the wave.

It is important to note that, the phase of the wave, $\omega(t - z/c)$ varies with both space and time, with c being the phase velocity of the wave. Moreover, it is evident from Eq. 3.18 that the wave propagates in space and time and has a definite well-defined period in time (with a period of T) and period in space (with a period of λ). This interpretation of plane wave brings in the following visual representation of a plane wave as shown in Fig. 3.3. This represents a wave propagation in space and time with a period of T and λ, respectively.

Another aspect is the interpretation of the solution of wave equation i.e, forward propagating plane wave, $f(x) = e^{-ikx}$. Mathematically, a shift of the function, $f(x)$ to the $+x$-direction produces the function, $f(x-a) = e^{-ik(x-a)}$. So, the function has shifted a distance $a = ct$ (say) over a time t beginning with $t = 0$. Thus, the function

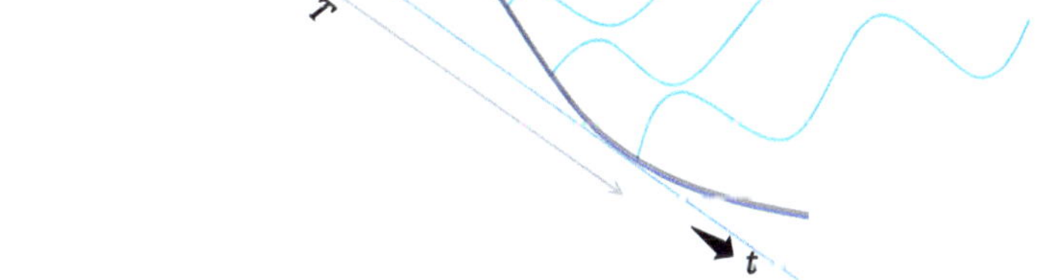

Fig. 3.3 A typical representation of a wave propagation in space and time

assumes the form, $f(x - ct) = e^{-ik(x-ct)} = e^{-i(kx-\omega t)}$, where $\omega = ck$. Noting that, $a = \omega t$, the solution conveys a wave shifted to an amount, $+a = \omega t$ after an elapsed time of t. This means that, $f(x - ct)$ represents a $forward - propagating\ wave$.

Note that, other important solutions of wave-equation do exists. These solution appear naturally when the wave equation is solved in other coordinate systems, such as, spherical coordinate or cylindrical coordinate. For example, a simple solution of wave equation in spherical coordinate system is a spherical wave. This is briefly discussed in the next sections. However, a complete understanding can be found in other textbooks [1,2].

3.3.3 Paraxial Helmholtz Equation

In general, the Helmholtz equation produces general solutions. It is particularly useful to obtain solutions in transverse XY plane (perpendicular to propagation direction Z). This will help realize a well-behaved beam (such as Gaussian beam) which is predominantly used in optical microscopes (confocal, two-photon and super-resolution).

The paraxial Helmholtz equation is associated with the transverse profile of the collimated beam. For simplicity, we will assume paraxial approximation, and define paraxial waves as the ones for which the wavefront normals are paraxial (parallel to the optical axis). Consider a paraxial plane wave propagating in z-direction,

$$\psi(r, z) = D(r)\, e^{-ikz} \tag{3.19}$$

This is a solution of the Helmholtz equation. To derive the transverse solution, we substitute the above equation (Eq. 3.19) in the Helmholtz equation, and separate out the transverse (x, y) and longitudinal (z) components,

$$\nabla_T^2(D\, e^{-ikz}) + \underbrace{\nabla_z^2(D\, e^{-ikz}) + k^2 D\, e^{-ikz}}_{G_z} = 0 \tag{3.20}$$

where, $\nabla_T^2 = \frac{\partial^2}{\partial x^2} + \frac{\partial^2}{\partial y^2}$ is the transverse Laplacian operator.

So, the transverse component is given by,

$$G_z = \nabla_z^2(D\, e^{-ikz}) - k^2 D e^{-ikz}$$

$$= \left(\frac{\partial^2 D}{\partial z^2} - i2k\frac{\partial D}{\partial z} + k^2 D\right)e^{-ikz} - k^2 D\, e^{-ikz} \tag{3.21}$$

To simplify the above equation we use slowly-varying envelop approximation which states that the variation in the envelop, $D(\vec{r})$ is slow within the distance $\Delta z = \lambda$. This immediately implies that, $\Delta D << D$. Using calculus, we can write,

$$\Delta D = \frac{\partial D}{\partial z}\,\Delta z \approx \frac{\partial D}{\partial z}\,\lambda \tag{3.22}$$

Hence, we have, $\frac{\Delta D}{\lambda} = \frac{\partial D}{\partial z}$, so that $\frac{\partial D}{\partial z} << \frac{D}{\lambda} = kD/2\pi$, giving,

$$\frac{\partial D}{\partial z} << kD \tag{3.23}$$

The above condition 3.23 implies,

$$\frac{\partial^2 D}{\partial z^2} << k\frac{\partial D}{\partial z} << k(kD) = k^2 D \tag{3.24}$$

Using the above inequality (Eq. 3.24), we ignore the first term in Eq. 3.21 and cancel the other term, which produces the following simplistic form G_z,

$$G_z = -i2k\frac{\partial D}{\partial z}\,e^{-ikz} \tag{3.25}$$

Incorporating G_z in Eq. 3.20 produces the following cohesive form for the paraxial Helmholtz equation,

$$\nabla_T^2 D - i2k\frac{\partial D}{\partial z} = 0. \tag{3.26}$$

The above equation (Eq. 3.26) is popularly known as the *paraxial Helmholtz equation*. This is a slowly-varying envelope approximation and is essentially the starting point for realizing a number of interesting solutions such as, Gaussian beam, Bessel beam, LG beams etc.

3.4 Gaussian Beams and their Properties

Now, we proceed to find the solution of *paraxial Helmholtz equation*. We will find that Gaussian beam turns out to be a natural solution. Since Gaussian beams are the most commonly used light sources in microscopy and imaging, so we will be discuss it in this section.

Consider a plane wave traveling along z-axis i.e,

$$\psi(x, y) = D(x, y)\,e^{-ikz} \tag{3.27}$$

We proceed to understand the behaviour of spherical wave (solution of Helmholtz equation in spherical coordinate system) under a slowly varying envelop approximation [2]. The spherical wave is given by,

$$D(\vec{r}) = D_0 \frac{e^{-ikr}}{r} \tag{3.28}$$

We further assume that the spherical waves originate at the origin ($\vec{r} = 0$) and seek solution far from the origin such that, $z >> \sqrt{x^2 + y^2}$. To determine the solution, we use Taylor series expansion of r i.e,

$$r = \sqrt{x^2 + y^2 + z^2} = z\sqrt{1 + \frac{x^2 + y^2}{z^2}} = z\left[1 + \frac{1}{2}\left(\frac{x^2 + y^2}{z^2}\right) + ...\right] \tag{3.29}$$

Retaining the first two terms and ignoring higher orders, we get,

$$r \approx z\left[1 + \frac{1}{2}\left(\frac{x^2 + y^2}{z^2}\right)\right] = z + \frac{x^2 + y^2}{2z}. \tag{3.30}$$

Incorporating the above expression (Eq. 3.30) in the spherical wave expression (Eq. 3.28), we get,

$$D(\vec{r}) \approx \frac{D_0}{z + (x^2 + y^2)/2z} e^{-ik\frac{x^2+y^2}{2z}} e^{-ikz} \tag{3.31}$$

Under the slowly varying envelop approximation, the magnitude remains nearly constant within a distance, $\Delta z = \lambda$, giving, $r = z + (x^2 + y^2)/2z \approx z$. So, the spherical wave can be expressed as,

$$D(\vec{r}) = D_0 \frac{e^{-ikz}}{z} e^{-ik\frac{x^2+y^2}{2z}} \tag{3.32}$$

Next, we generalize the spherical wave solution by substituting $z \rightarrow s(z) = z - \zeta$, and take ζ as purely complex, $\zeta = -iz_0$. Incorporating these changes, we get,

$$D(\vec{r}) = D_0 \frac{e^{-iks(z)}}{s(z)} e^{-ik\frac{x^2+y^2}{2s(z)}} \tag{3.33}$$

where, $s(z) = z + iz_0$.

Finally, we substitute, $\frac{1}{s(z)} = \frac{1}{R(z)} + (i\frac{\lambda}{\pi})\frac{1}{W^2(z)}$ to separate out the real and complex terms. This produces the generalized solution,

$$\psi = D_0' \left[\frac{W_0}{W(z)}\right] e^{-(x^2+y^2)/W^2(z)} \exp\left[-i\left(k\frac{(x^2 + y^2)}{2R(z)} + \Xi(z) - kz\right)\right] \tag{3.34}$$

where, D_0' is the cumulative constant. Other parameters are,

$$\begin{cases} W_0 = \sqrt{\frac{z_0\lambda}{\pi}} & beam-waist\ at\ z_0 \\ W(z) = \sqrt{1+\frac{z}{z_0}}\, W_0 & beam-width\ at\ z \\ R(z) = z\left[1+\frac{z_0^2}{z^2}\right] & radius\ of\ curvature \\ \Xi = \tan^{-1}\left(\frac{z}{z_0}\right) & Gouy\ phase \end{cases} \tag{3.35}$$

The above equation (Eq. 3.34) is a general solution of *paraxial Helmholtz equation*, also known as *paraboloidal wave*.

3.4.1 Gaussian Beams

The general solution is the starting point for several solutions of paraxial Helmoltz equation, of which Gaussian beam is one. Here, we are interested in Gaussian beam and study its profile in the transverse plane.

Since we are interested in the transverse profile intensity, we take modulus of the general solution (ψ) i.e.,

$$I(\rho, z) = |\psi|^2 = I_0\left[\frac{W_0}{W(z)}\right]^2 e^{-\frac{2\rho^2}{W^2(z)}} \tag{3.36}$$

where, $I_0 = |D_0|^2$ is a constant.

The above equation represents a Gaussian beam at any fixed z-position. Along the longitudinal direction, it is maximum at $z = 0$. However, the beam has maximum at $\rho = \sqrt{x^2+y^2} = 0$ which is, $I(\rho, z = 0) = D_0$, and the intensity decreases monotonically far from it ($\rho \to \infty$). Figure 3.4 gives a representative image of Gaussian

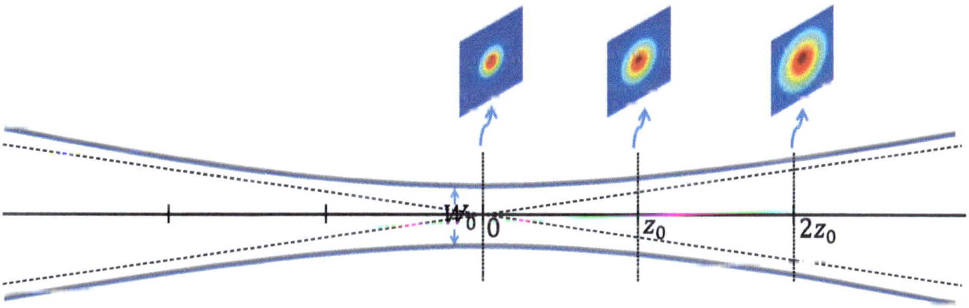

Fig. 3.4 Gaussian beam as a solution of *paraxial Helmholtz equation* with its profile at $z = 0,\ z_0,\ 2z_0$

profile of the beam at, $z = 0,\ z_0, 2z_0$. However, the maximum of the Gaussian beam i.e, $I_{max} = I(0, 0) = D_0$ occurs at $(\rho = 0,\ z = 0)$.

3.4.2 Properties of Gaussian Beams

Next we investigate the properties of Gaussian beam as it propagates along the z-axis.

Power (P) The power of the beam is the integration of intensity in the area of radius ρ i.e,

$$P = \int_0^{\infty} I(\rho, z)\, 2\pi\rho\, d\rho \tag{3.37}$$

Noting that the intensity at $\rho = 0$ is I_0, the integration yields,

$$P = I_0\, (\pi W_0^2)/2 \tag{3.38}$$

Phase (φ) From Eq. 3.34, the complete phase of the wavefunction is,

$$\varphi = k\frac{(x^2 + y^2)}{2R(z)} + \Xi(z) - kz \tag{3.39}$$

Note that, for $\rho = 0$ (on the axis), the phase is,

$$\varphi = kz - \Xi(z) \tag{3.40}$$

Here, kz is the phase collected by the plane wave over a distance z, and the second term, Ξ is phase retardation (known as *Gouy phase*). The total Gouy phase accumulated by the wave is π (from $-\pi/2$ to $+\pi/2$) as the wave travels from $-\infty$ to $+\infty$. This is the phase retardation due to the delay of the wavefront with respect to spherical wave.

Beam Width, $W(z)$ This refers to the width of the beam at any z as it goes from $-\infty$ to $+\infty$. In specific, the minimum beam-width is, W_0 at $z = 0$. Explicitly, the beam-width is given by,

$$W(z) = \sqrt{1 + \frac{z^2}{z_0^2}}\, W_0 = \sqrt{1 + \frac{z^2}{z_0^2}}\sqrt{\frac{z_0\lambda}{\pi}} \tag{3.41}$$

Figure 3.4 shows the shape and intensity of the beam at few selected z-points ($z = 0,\ z_0,\ 2z_0$).

Depth of Focus or Rayleigh Range, D_f This is defined as the distance at which the beam-waist falls off by a factor of $\sqrt{2}$ or equivalently the intensity decreases by a factor of 2 on either side of $z = 0$, i.e,

$$D_f = 2z_0 = \frac{2\pi}{\lambda} W_0^2 \tag{3.42}$$

This states that, the DOF is directly proportional to the square of beam-width or equivalently, the area of the beam πW_0^2. Both beam-width and DOF have prominent roles in microscopy as it determines the spot-size, and penetration-depth.

Although imaging systems often use Gaussian beams for excitation, there exists other kinds of beams such as, Bessel beam, Laguerre-Gaussian (LG) beam, and doughnut beam. These specialized beam shapes have special characteristics/intensity profiles (large depth-of-focus or zero central intensity) which can be used for specific applications in optical microscopy. We refer readers to dedicated literature to explore these specialized beams [2, 3].

3.5 Diffraction of Light

In standard terms, diffraction of light refers to the wave-like properties of light and its spread as it propagates through an edge or medium or optical element (e.g., lens). One example, where diffraction is apparent is light passing through a circular aperture. This does not merely cast the aperture's shadow but has additional features such as concentric circles. Looking at the situation, it appears that light behaves as a wave, bends at the edges of the aperture, and propagates to the screen, where it forms the image of the aperture. Thus, the image formed has distinct characteristics that can be explained based on the diffraction of light and transfer function. In this section, we will discuss this particular nature of light and its effect leading to the optical resolution of an optical microscopy system.

3.5.1 Transfer Function in Free Space

Consider a plane wave propagating from $z = 0$ to $z = d$. The complex amplitude at the input and output planes are, $f(x, y) = \psi(x, y, z = 0)$ and $g(x, y) = \psi(x, y, z = d)$, respectively (see, Fig. 3.5). The propagation of light from input to output plane is characterized by an impulse response, $h(x, y)$ or equivalently by the transfer function, $H(\nu_x, \nu_y)$. We assume the whole process to be a linear shift-invariant system fully characterized by the transfer function. So, the relation between the input and output function is given by,

$$\begin{cases} g(x, y) = h(x, y) \otimes f(x, y) & \textit{space domain} \\ G(\nu_x, \nu_y) = H(\nu_x, \nu_y)\, F(\nu_x, \nu_y) & \textit{frequency domain} \end{cases} \tag{3.43}$$

where, $\otimes$ represents convolution operation.

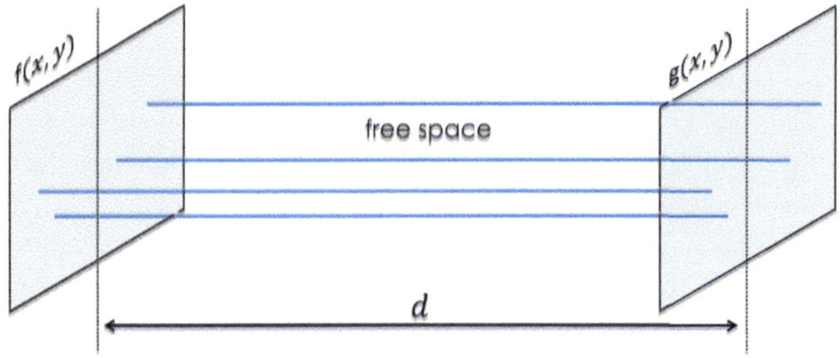

Fig. 3.5 A graphical representation of wave propagation from $z = 0$ to d

Taking the input plane wave as, $f(x, y) = A\ e^{-j(k_x x + k_y y)}$ for which the output at a distance d is given by, $g(x, y) = A\ e^{-j(k_x x + k_y y + k_z d)}$, where, $k_x = 2\pi\nu_x$, $k_y = 2\pi\nu_y$, and $k_z = \sqrt{k^2 - k_x^2 - k_y^2}$. The corresponding transfer function is given by,

$$H(\nu_x, \nu_y) = \frac{G(\nu_x, \nu_y)}{F(\nu_x, \nu_y)} = e^{-jk_z d} \tag{3.44}$$

Using the relation, $k_z = \sqrt{k^2 - (k_x^2 + k_y^2)}$ in Eq. 3.44 produces,

$$\begin{aligned} H(\nu_x, \nu_y) &= \exp[-j\sqrt{k^2 - (k_x^2 + k_y^2)}\ d\ ] \\ &= \exp\left[-j2\pi d\sqrt{\frac{1}{\lambda^2} - (\nu_x^2 + \nu_y^2)}\right] \end{aligned} \tag{3.45}$$

where, $k = 2\pi/\lambda$, $\nu_x = k_x/2\pi$, and $\nu_y = k_y/2\pi$.

The above equation (Eq. 3.45) is the transfer function for a free space propagation by light. Note that, the attenuation factor has a cutoff of λ^{-1}, which is the spatial bandwidth. Hence, features having spatial frequencies greater than λ^{-1} (corresponding to details finer than λ) cannot be transmitted by a light wave over long distance.

3.5.2 Fresnel Diffraction

Fresnel diffraction is a consequence of Fresnel approximation which is equivalent to Fourier transform in the near field. So, we focus our attention on the transfer function in the near field. Accordingly, we consider the following assumption i.e.,

> The input function $f(x, y)$ contain spatial frequencies (ν_x, ν_y) that are much smaller than the cutoff frequency $(1/\lambda)$.

We use the above assumption to simplify the transfer function (Eq. 3.45). This can be accomplished by expanding and reducing the phase term i.e.,

$$\phi = 2\pi d\sqrt{\frac{1}{\lambda^2} - (\nu_x^2 + \nu_y^2)} \tag{3.46}$$

We use the Fresnel approximation (i.e., $\nu_x^2 + \nu_y^2 << 1/\lambda^2$) to the phase term (Eq. 3.46) and use Binomial expansion giving,

$$\phi = \frac{2\pi d}{\lambda}\left[1 - \lambda^2\nu_T^2\right]^{1/2} = 2\pi\frac{d}{\lambda}\left[1 - \frac{\lambda^2\nu_T^2}{2} + ...\right] \tag{3.47}$$

where, $\nu_T = \sqrt{\nu_x^2 + \nu_y^2}$.

Considering the first two terms and ignoring higher orders, the transfer function for free-space becomes,

$$H(\nu_x, \nu_y) = H_0\, e^{-j\pi d\lambda\nu_T^2} \tag{3.48}$$

where, $H_0 = e^{-jkd}$.

This is known as the **Fresnel approximation** that forms the basis for Fresnel diffraction (diffraction at the near-field) as discussed in the next section.

Note that, the Fresnel approximation in the Fourier domain (ν_x, ν_y) resembles a 2D Gaussian function (see, Eq. 3.48). Since, the Fourier transform of a 2D Gaussian is a 2D Gaussian function, the same (Eq. 3.48) can be expressed in the spatial domain. So, taking inverse 2D Fourier transform, we get,

$$h(x, y) = \frac{j}{\lambda d}\, e^{-jk(x^2+y^2)/2d} \tag{3.49}$$

We will use both the relations (3.48 in frequency domain and 3.49 in space domain) effectively throughout the text book.

Condition of Validity The validity of Fresnel approximation is based on the condition that the third term is much less than π for all θ i.e,

$$2\pi\frac{d}{\lambda}\frac{\theta^4}{8} << \pi \quad \Rightarrow \frac{\theta^4 d}{4\lambda} << 1 \tag{3.50}$$

where, $\theta_x \approx \lambda\nu_x$; $\theta_y \approx \lambda\nu_y$, and $\theta^2 = \lambda^2(\theta_x^2 + \theta_y^2)$ is the angle subtended by largest radial distance in the output plane.

Considering a as the largest radial distance, and d as the distance between the input and output plane, the angle $\theta_{max} \approx a/d$. Incorporating this in the above equation produces,

$$N_F \frac{\theta_{max}^2}{4} << 1 \tag{3.51}$$

where,

$$N_F = \frac{a^2}{\lambda d} \tag{3.52}$$

is the Fresnel number.

Here, the Fresnel number is defined to quantify the ratio between the area in the output plane $\sim a^2$ and the distance d. The ratio has special significance in determining the diffraction pattern. This also enhances the fact that diffraction pattern is strongly influenced by distance and area.

3.5.3 Free Space Propagation as a Convolution Operation

To determine the relation between the input and output plane in space domain, we use superposition of discrete points/delta functions originating in the input plane $f(x, y)$. Here each point may be considered as the source producing a spherical wave. These waves travel from input to output plane and the change is characterized by the impulse response, $h(x, y)$. So, the wave originating at a point (x_1, y_1) in the input plane has an amplitude, $f(x_1, y_1)$. It travels a distance d before it reaches the output plane. In the process, the wave gets modified by the impulse response, giving the amplitude in the output plane as, $f(x_1, y_1) \times h(x - x_1,\ y - y_1)$. Since the contribution at a point (x, y) is from all the points in the input plane, so the amplitude in the output plane is given by,

$$\begin{aligned} g(x, y) &= \int_{-\infty}^{+\infty} \int_{-\infty}^{+\infty} f(x_1, y_1) \times h(x - x_1,\ y - y_1)\, dx_1\, dy_1 \\ &= f(x, y) \bigotimes h(x, y) \end{aligned} \tag{3.53}$$

So, the output is a convolution of input function and the impulse response that characterizes propagation of light in free space.

Further expansion of the above expression is possible by engaging Fresnel approximation 3.49, giving,

$$g(x, y) = h_0 \int_{-\infty}^{+\infty} \int_{-\infty}^{+\infty} f(x_1, y_1)\ \exp\left\{-j\pi \frac{(x - x_1)^2 + (y - y_1)^2}{\lambda d}\right\} dx_1\, dy_1 \tag{3.54}$$

where, $h_0 = \frac{j}{kd}\, e^{-jkd}$ that determines the phase collected by the wave for a propagation distance of d.

It is important to realize that the above expression is a general expression and can be used throughout the space.

3.5.4 Fraunhofer Approximation

Another important consideration is that of far-field effects of diffraction from input to output plane. In this case, Fraunhoher approximation is applicable, that states,

> At long distances, one only needs to consider those waves contributing to the complex amplitude at a point in the output plane that makes small angles with the optical axis, such that $(\theta_x \approx x/d,\ \theta_y \approx y/d)$. The contribution from all other waves cancels out due to destructive interference.

Using the Fraunhofer approximation, the wavevector components can be expressed as,

$$\begin{cases} k_x = k\,\theta_x \approx k(x/d) \\ k_y = k\,\theta_y \approx k(y/d) \end{cases} \tag{3.55}$$

Noting that, the complex amplitude in the output plane is given by,

$$g(x, y) = h_0 \int_{-\infty}^{+\infty}\int_{-\infty}^{+\infty} f(x_1, y_1)\ \exp\left\{-j\pi \frac{(x - x_1)^2 + (y - y_1)^2}{\lambda d}\right\} dx_1\, dy_1 \tag{3.56}$$

for which the phase is,

$$\varphi(x, y) = \frac{\pi}{\lambda d}[(x - x_1)^2 + (y - y_1)^2] \tag{3.57}$$

Expanding $\varphi(x, y)$ we get,

$$\varphi(x, y) = \frac{\pi}{\lambda}\left[\frac{(x^2 + y^2)}{d} + \frac{(x_1^2 + y_1^2)}{d} - \frac{2(xx_1 + yy_1)}{d}\right] \tag{3.58}$$

Now we use Fraunhofer approximation which ensures that the aperture of radius for the output function, g and input function f i.e, $r = \sqrt{x^2 + y^2}$ and $r_1 = \sqrt{x_1^2 + y_1^2}$ are very small comparted to the distance, d between them (small angles). So, the first and second term in the above Eq. 3.58 can be neglected, producing the following simplified form,

$$\varphi(x, y) \approx -\frac{2\pi}{\lambda d}(xx_1 + yy_1) \tag{3.59}$$

It may be noted that, the condition imposed by Fraunhofer approximation i.e, $r,\ r_1 << d$ leads to,

$$\begin{cases} N_F = \frac{r^2}{\lambda d} << 1 \\ N_{1F} = \frac{r_1^2}{\lambda d} << 1 \end{cases} \tag{3.60}$$

Substituting Eq. 3.59 in Eq. 3.56, gives the following expression for the output function,

$$\begin{aligned} g(x, y) &\approx h_0 \int\int_{-\infty}^{+\infty} f(x_1, y_1) e^{-j\varphi(x,y)}\, dx_1\, dy_1 \\ &= h_0 \int\int_{-\infty}^{+\infty} f(x_1, y_1) \exp\left[-j\frac{\pi}{\lambda d}(xx_1 + yy_1)\right] dx_1\, dy_1 \\ &= h_0\, F(x/\lambda d,\ y/\lambda d) \\ &= h_0\, F(k_x/2\pi,\ k_y/2\pi) \\ &= h_0\, F(\nu_x,\ \nu_y) \end{aligned} \tag{3.61}$$

where, $k_x = 2\pi\nu_x$ and $k_y = 2\pi\nu_y$, and $h_0 = \frac{j}{\lambda d}\, e^{-jkd}$. Note that, $F(...)$ represents Fourier transform.

Equation 3.61 clearly states that the complex amplitude in the output plane is the product of complex quantity h_0 and the Fourier transform of the input function (evaluated at spatial frequencies, ν_x and ν_y) in the **Fraunhofer approximation**.

Summarizing, the approximations used for determining the diffraction pattern at small and large distances are different, with Fresnel approximation dictating the diffraction pattern at small distances, while Fraunhofer approximation is applicable at large distances (distances larger than wavelength of light). In the subsequent sections, we will use these approximations to determine diffraction pattern at the focus of objective lens. This will help us determine the system resolution of optical microscopes.

3.6 Fresnel and Fraunhofer Diffraction

It is important to realize that diffraction pattern in the output plane depends on its distance from the input plane. Fresnel's approximation is the most accurate for short distances, whereas Fraunhofer approximation is applicable at long distances. Alternately, this can also be understood based on Fresnel number (see, Eq. 3.52). For small Fresnel number, Fraunhofer approximation accurately determines the diffraction whereas Fresnel approximation suits better at large Fresnel number. Going forward, we will discuss and determine the mathematical expression for both types of diffraction.

We begin with diffraction at short distances i.e, very close to the input plane. To determine the diffraction pattern, the input complex amplitude (say, $\psi(x, y)$) is allowed to pass through an aperture (with aperture function, $\eta(x, y)$) and its free-space propagation is determined by Fresnel approximation.

Assuming an unit function, $\psi(x, y) = 1$ for the incident wave, the function just after the aperture is the product of unit function for amplitude and aperture function i.e, $f(x, y) = \eta(x, y)$. The modified complex amplitude then travels a distance of d before reaching the output plane. This gives the output amplitude function as,

$$g^{(fres)}(x, y) = h_0 \int\int \eta(x, y)\, exp\left[-j\pi \frac{(x - x_1)^2 + (y - y_1)^2}{\lambda d}\right] dx_1\, dy_1 \quad (3.62)$$

The diffraction intensity pattern at a distance d is,

$$\begin{aligned} I^{(fres)}(x, y) &= |g^{(fres)}(x, y)|^2 \\ &= \frac{1}{\lambda d}\left| \int\int \eta(x, y)\, exp\left[-j\pi \frac{(x - x_1)^2 + (y - y_1)^2}{\lambda d}\right] dx_1\, dy_1 \right|^2 \end{aligned} \quad (3.63)$$

This is a general expression for determining diffraction pattern at small distances, termed as, **Fresnel diffraction**.

However, one can use Fraunhofer approximation for determining the effect of wave propagation in free-space at long distances. Similar to the previous case, the input complex amplitude of the wave pass through the aperture and thereafter the free-space propagation is given by Fraunhofer approximation. This gives the following complex amplitude at the output plane (placed at large distance, d),

$$g(x, y) \approx h_0\, F\left(\frac{x}{\lambda d}, \frac{y}{\lambda d}\right) \quad (3.64)$$

where, $h_0 = \frac{j}{\lambda d}\, e^{-jkd}$.

Noting that, the input function is $f(x, y) = \psi(x, y) \times \eta(x, y) = \eta(x, y)$, the above equation becomes,

$$g^{fraun}(x, y) \approx h_0\, \Xi\left(\frac{x}{\lambda d}, \frac{y}{\lambda d}\right) \quad (3.65)$$

where, Ξ is the Fourier transform of $\eta(x, y)$ i.e,

$$\Xi\left(\frac{x}{\lambda d}, \frac{y}{\lambda d}\right) = \Xi(\nu_x, \nu_y) = \int\int \eta(x, y)\, e^{-j2\pi(\nu_x x + \nu_y y)}\, dx\, dy \quad (3.66)$$

So, the diffraction pattern at large distances (known as, **Fraunhofer diffraction**) is given by,

$$I^{fraun} = |g^{fraun}(x, y)|^2 = \frac{1}{\lambda d} \left| \Xi(\nu_x, \nu_y) \right|^2 \tag{3.67}$$

The above expression states that, the diffraction pattern at large distances is the Fourier transform of the aperture function (weighted by wave-amplitude at each point in the aperture).

3.7 Diffraction from a Circular Aperture

In this section, we will take the example of circular aperture to understand diffraction pattern in the far-field (i.e, Fraunhofer diffraction). As usual, we assume uniform unit illumination at the back of the aperture and the propagation of the light wave from the aperture to the output plane to be described by Fraunhofer approximation. This condition prevails in most of the optical microscopy systems.

To begin with, the circular aperture can be expressed as,

$$\eta(x, y) = \begin{cases} 1, & r = \sqrt{x^2 + y^2} < d_0/2 \\ 0, & r = \sqrt{x^2 + y^2} > d_0/2 \end{cases} \tag{3.68}$$

Considering uniform unit illumination, $\psi(x, y) = 1$, the complex amplitude just after the aperture is,

$$f(x, y) = \psi(x, y) \times \eta(x, y) = \eta(x, y) \tag{3.69}$$

Under Fraunhofer approximation (see, Eq. 3.61, the complex amplitude at the output plane is given by,

$$g(x, y) = h_0 \, \Xi(\nu_x, \nu_y) \tag{3.70}$$

where, $\nu_x = \frac{x}{\lambda D}$ and $\nu_y = \frac{y}{\lambda D}$. Here, Ξ is the Fourier transform of the aperture function,

$$\Xi(\nu_x, \nu_y) = \int\int \eta(x, y) \, e^{j2\pi(\nu_x x + \nu_y y)} \, d\nu_x \, d\nu_y \tag{3.71}$$

So, the diffraction pattern in the far-field of a circular aperture is given by,

$$I(x, y) = |g(x, y)|^2 = \frac{1}{\lambda d} |\Xi(\nu_x, \nu_y)|^2 \tag{3.72}$$

Now, we need to find the Fourier transform of the aperture function. Due to the circular symmetry of the aperture, we perform Fourier transform in the polar coordinate system. Accordingly, we make the following transformation both in spatial and Fourier domains,

$$\begin{cases} x = r\cos\theta \\ y = r\sin\theta \end{cases} \Rightarrow \begin{cases} r = \sqrt{x^2+y^2} \\ \theta = \tan^{-1}(y/x) \end{cases}$$

$$\begin{cases} \nu_x = \rho\cos\phi \\ \nu_y = \rho\sin\phi \end{cases} \Rightarrow \begin{cases} \rho = \sqrt{\nu_x^2+\nu_y^2} \\ \theta = \tan^{-1}(\nu_y/\nu_x) \end{cases}$$

Using the above transformation, the Fourier transform in the polar coordinate system produces,

$$\begin{aligned} \Xi(\rho,\phi) &= \int_{r=0}^{\infty}\int_{\theta=0}^{2\pi} \eta(r,\theta)\, e^{j2\pi r\rho(\cos\theta\cos\phi-\sin\theta\sin\phi)}\, r\,dr\,d\theta \\ &= \int_{r=0}^{d_0/2}\left[\int_{\theta=0}^{2\pi} e^{j2\pi\rho\cos(\theta-\phi)}d\theta\right] r\,dr \end{aligned} \tag{3.73}$$

Let, $\xi - 2\pi r\rho$ hence, $d\xi = 2\pi\rho dr$. Substitution gives the following compact form,

$$\begin{aligned} \Xi(\rho,\phi) &= \frac{1}{2\pi\rho^2}\int_{\xi=0}^{\phi\rho d_0} \underbrace{\left[\frac{1}{2\pi}\int_{\theta=0}^{2\pi} e^{j\xi\cos(\theta-\phi)}\,d\theta\right]}_{J_0} \xi\,d\xi \\ &= \frac{1}{2\pi\rho^2}\int_{\xi=0}^{\pi\rho d_0} \xi\, J_0(\xi)\,d\xi \end{aligned} \tag{3.74}$$

where, J_0 is a zero-order Bessel function.

Using the Bessel identity, $\int_0^a x\, J_0(x)\,dx = a\, J_1(a)$, and substituting, $\rho = \sqrt{\nu_x^2+\nu_y^2} = r/\lambda d_0$, we get,

$$\Xi(\rho,\phi) = \frac{1}{2\pi\rho^2}\,(\pi\rho d_0)\, J_1(\pi\rho d_0) \tag{3.75}$$

The diffraction pattern at the far-field in the spatial domain is,

$$I(x,y) = \frac{\pi d_0^2}{(\lambda D)^2}\left|\frac{J_1(\chi)}{\chi}\right|^2 \tag{3.76}$$

where, $\chi = \pi r d_0/\lambda D$.

So, the diffraction pattern of a circular aperture at the far-field is modulus-square of the $J(\chi)/\chi$ function. The diffraction pattern is as shown in Fig. 3.6.

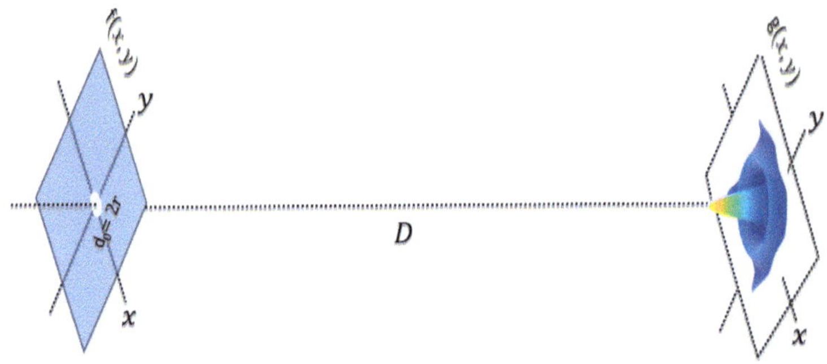

Fig. 3.6 Diffraction from a circular aperture (size ∼ d_0) on a screen placed at a distance D (Fraunhofer Diffraction)

3.8 Optical Resolution

The optical resolution of a microscopy system depends on the focusing element i.e, objective lens. Assuming that maximum aperture of the lens is used, and the beam is tightly focused to a small spot. This spot of light is essentially the diffraction pattern of the light at far-field. A schematic diagram depicting this situation is shown in Fig. 3.7. A lens of size d_0 is exposed while the rest of light is blocked. Thereafter, the complex amplitude of the light wave propagates a long focal-distance $D = f$ before it falls on the output plane. Since the diffraction pattern is sought at large distance, the complex amplitude in the far-field is dictated by Fraunhofer approximation.

We can directly use the expression for diffraction intensity pattern (Eq. 3.76) for a circular aperture (size $d_0 = 2r$) at the focus ($D = f$) of the lens i.e,

$$I(x, y) = \frac{\pi d_0^2}{(\lambda f)^2} \left| 2\frac{J_1(\chi)}{\chi} \right|^2 \tag{3.77}$$

where, $\chi = \pi d_0 \rho / \lambda f$.

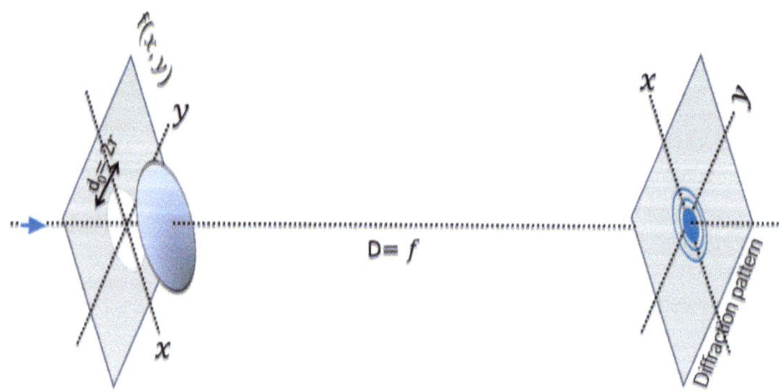

Fig. 3.7 Diffraction from a circular aperture and focusing lens on a screen placed at a distance D

The most important function here is the $|J_1(\chi)/\chi|^2$ that resembles a central lobe encircled by multiple rings of decreasing intensity. The shape can be verified using Matlab scripts. A typical such function is shown in Fig. 3.7. An interesting question to ask at this stage is, "*How close two such functions can be and still be distinguishable as separate entities.*" This brings-in the notion of resolution of an optical microscopy system (specifically, the objective lens). An approximate way to answering this question would be to identify the maximum contrast point as we bring two such diffraction patterns closer and closer. From the figure, it is clearly evident that, the overlap of the maximum of the central spot of first diffraction spot with the first minimum of the second diffraction spot would give the maximum contrast. Closer than this would merge both the diffraction spots, making them indistinguishable. Thus, the closest the diffraction patterns can be to remain distinguishable can be defined as the first minimum distance from its center.

Taking this as the guiding principle, the resolution of a microscopy system (with objective as the focusing element) is the first minima (say, r_1) of the diffraction pattern. At this point the intensity is near zero i.e.,

$$I(x,y)\,|_{r_1} = 0 \quad \Rightarrow J_1(\chi)\,|_{r_1} = 0 \tag{3.78}$$

Now, we use standard table of Bessel function to find its first zero which occurs at 3.8317, giving,

$$\chi\,|_{r_1} = \frac{\pi r_1 d_0}{\lambda f} = 3.8317 \quad (first\ minimum)$$

$$\Rightarrow r_1 = \frac{1.22\,\lambda f}{d_0} \tag{3.79}$$

Hence, the minimum resolvable distance between two diffraction spots is,

$$\zeta = r_1 = \frac{1.22\lambda}{2(\frac{d_0/2}{f})} \sim \frac{1.22\lambda}{2\sin\alpha} \tag{3.80}$$

where, $d_0/2$ is the radial distance of the lens aperture, f is the focal length, and α is the aperture angle.

In general, modern optical microscopes employ high NA objectives for focusing light. This calls for water or oil immersion objectives. Light passing through objective immersion medium and then in the sample results in wavelength scaling i.e, $\lambda \rightarrow \lambda_0/n$, where n is the refractive index of the immersion medium or sample and λ_0 is the wavelength of light in air. Finally, this leads to the following expression for optical resolution,

$$\zeta = \frac{1.22\,\lambda_0}{2n\sin\alpha} \tag{3.81}$$

The above relation is the classical diffraction limit and also known as the **Rayleigh criterion (1.22) or Abbe's diffraction limit (1)**. Note that for Abbe limit, 1.22 is replaced by 1.

Irrespective of the nature or design of classical imaging system, the classical resolution is bound by the above relation 3.81. In the the subsequent chapters, we will use this relation repetidely and find ways to overcome the diffraction limit which will lead to super-resolution in optical microscopy.

References

1. Mondal PP (2021) Classical and quantum optics. American Institute of Physics Publisher, New York
2. Saleh BEA, Teich MC (2012) Fundamentals of Photonics, 2nd edn. Wiley, New York
3. Mondal PP (2021) Classical and Quantum Optics. American Institute of Physics publisher, New York

4 Electromagnetic Optics, Light-Matter Interaction and Optical Microscopy

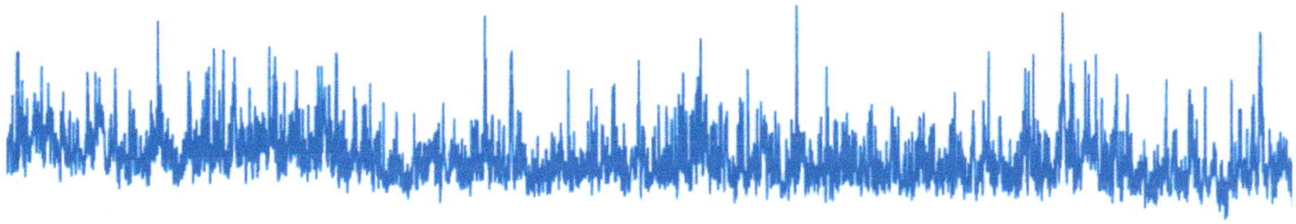

Although ray and wave optics are capable of explaining most of the day-to-day observations, they are incapable of explaining many important effects of light. These effects, although on small scales give rise to a noticeable outcomes. Some of the examples include, electric effects of light on certain metal surfaces (plasmonics), and light passing through birefringent materials and specific crystals (such as, Quartz). In addition, light gets altered by magnetic fields causing it to rotate when it passes through certain media in the presence of a magnetic field (Faraday effect). These and many other examples prompt us to believe that light does have non-zero electric and magnetic effects. So, light can be further subclassified as magnetic and electric field vector waves also known as electromagnetic waves. Concentrated electromagnetic fields are an essential feature of optical microscopy and high NA objective lenses are used to tightly focus electric fields. The theory allows computation of electric field at the focus which forms the basis to understand all kinds of microscopy systems (epifluorescence, confocal and two-photon microscopy). In addition, field-dipole interaction is the foundation for all kinds of polarization microscopy. This gives the essence of the electromagnetic nature of light in optical microscopy.

Understanding the electromagnetic nature of light and its effects is of utmost importance in optical microscopy. In fact, the study of electromagnetic fields is central to realize high-quality imaging and precision microscopy. Over the years, several experiments have shown the electric and magnetic effects of light. One of the early experiments is related to birefringent materials such as quartz that causes phase change of EM waves. For

P. P. Mondal, S. Hess, *Single-Molecule Super-Resolution Optical Microscopy*,
https://doi.org/10.1007/978-3-032-13442-4_4

bifrindgent materials, the refractive index varies depending on the electric field direction (polarization) of light. This means the velocity changes, which leads to phase difference, resulting in an altered polarization state of the transmitted light. In another example, the fraction of light reflected or transmitted is found to depend on the polarization of incident light. Recent advances have shown the interaction of electromagnetic fields with electrons at the surface of a metal, which gives rise to collective oscillations of electrons. These oscillations are known as surface plasmons or simply plasmons. This effect is heavily used in surface plasmon near field microscopy for nanoscale imaging. Thus plasmonics show the effect of electric field of light on matter. All the three examples go on to suggest the strong presence of electric and magnetic effects of light.

In this chapter, we will discuss some of these effects based on a concrete theoretical formulation of electromagnetic fields. Specifically, we will elaborate on the origin of the EM wave equation, and possible solutions; the polarization of light and the devices based on it; angular spectrum representation necessary for field distribution at the focus of objective lens; some of the well-known fluorescence microscopy systems; and field-dipole interactions in optical microscopy.

4.1 Wave Equation and Its Solution

We begin our discussion with the basic laws of electrostatics and magnetostatics in differential form, and show how this set of laws gives rise to the EM wave equation. To do so, we assume that light is electromagnetic in nature and the EM field is described by two related vector fields: *Electric field,* $\vec{E}(\vec{r}, t)$ and *Magnetic field,* $\vec{H}(\vec{r}, t)$. Note that, the electric and magnetic nature of light is encapsuled in Maxwell's equation which intern is a differential representation of well-known laws in electrostatic and magnetostatics.

So, we begin with the Maxwell's equations in a region without charges and currents i.e.,

$$\begin{cases} \vec{\nabla}.\vec{E} = 0 \\ \vec{\nabla}.\vec{H} = 0 \\ \vec{\nabla} \times \vec{E} = -\mu_0 \frac{\partial \vec{H}}{\partial t} \\ \vec{\nabla} \times \vec{H} = \epsilon_0 \frac{\partial \vec{E}}{\partial t} \end{cases} \tag{4.1}$$

where, $\epsilon_0 = \frac{1}{36\pi} \times 10^{-9}\ F/m$ is electric permittivity which is the measure of electric field effects on the medium, and $\mu_0 = 4\pi \times 10^{-7}$ is the magnetic permeability which is the ability of a medium (here, air) to support the formation of magnetic field in it when exposed to external magnetic field.

Next, we proceed to derive electromagnetic wave equation from Maxwell's equations 4.1. To do so, we take *Curl* operation on third equation i.e,

$$\vec{\nabla} \times (\vec{\nabla} \times \vec{E}) = -\mu_0 \frac{\partial}{\partial t}(\vec{\nabla} \times \vec{H}) \tag{4.2}$$

Expanding left-hand side and using fourth equation from 4.1 we get,

$$\vec{\nabla}(\vec{\nabla}.\vec{E}) - \vec{\nabla}.(\vec{\nabla}\vec{E}) = -\mu_0\epsilon_0 \frac{\partial^2 \vec{E}}{\partial t^2} \tag{4.3}$$

Using first Maxwell's equation and noting that, $\vec{\nabla}.(\vec{\nabla}\vec{E}) = \vec{\nabla}^2\vec{E}$, produces,

$$\vec{\nabla}^2\vec{E} - \frac{1}{c_0^2}\frac{\partial^2 \vec{E}}{\partial t^2} = 0 \tag{4.4}$$

where, $c_0 = 1/\sqrt{\epsilon_0\mu_0}$ is the speed of light in the medium (here, vacuum).[1]

We seek the solution to the EM wave equation 4.4. More specifically, we seek the solution for the spatial part only i.e,

$$\vec{\nabla}^2\vec{E} = -k^2\,\vec{E} \tag{4.5}$$

where, $k^2 = \omega^2\mu_0\epsilon_0$.[2]

In the Cartesian coordinate system, the field vector $\vec{E}$ can be expanded as,

$$\vec{E}(x, y, z) = E_x(x, y, z)\,\hat{x} + E_y(x, y, z)\,\hat{y} + E_z(x, y, z)\,\hat{z} \tag{4.6}$$

Substituting Eq. 4.6 in the wave equation (Eq. 4.5) and separating component-wise, produces a set of three separate equations,

$$\begin{cases} \vec{\nabla}^2.E_x(x, y, z) + k^2\,E_x(x, y, z) = 0 \\ \vec{\nabla}^2.E_y(x, y, z) + k^2\,E_y(x, y, z) = 0 \\ \vec{\nabla}^2.E_z(x, y, z) + k^2\,E_z(x, y, z) = 0 \end{cases} \tag{4.7}$$

Since all the components have similar form, so solving one of them is enough. Accordingly, we take wave-equation corresponding to E_x components and use separation-

[1] Note that, a similar equation exists for magnetic field vector i.e, $\vec{\nabla}^2\vec{H} - \frac{1}{c_0^2}\frac{\partial^2\vec{H}}{\partial t} = 0$

[2] Note that, this can be obtained from the time-varying field by assuming time variation of the form, $e^{i\omega t}$, and by replacing $\frac{\partial}{\partial t}$ by $i\omega$.

of-variables method i.e,

$$E_x(x, y, z) = u(x) \times v(y) \times w(z) \tag{4.8}$$

Substituting, the above solution in Eq. 4.7, and dividing by $u \times v \times w$ gives,

$$\frac{1}{u}\frac{\partial^2 u}{\partial x^2} + \frac{1}{v}\frac{\partial^2 v}{\partial y^2} + \frac{1}{w}\frac{\partial^2 w}{\partial z^2} = -k^2 \tag{4.9}$$

Expressing, $k^2 = k_x^2 + k_y^2 + k_z^2$ and separating each term produces,

$$\begin{cases} \frac{1}{u}\frac{\partial^2 u}{\partial x^2} + k_x^2 u = 0 \\ \frac{1}{v}\frac{\partial^2 v}{\partial y^2} + k_y^2 v = 0 \\ \frac{1}{w}\frac{\partial^2 w}{\partial x^2} + k_z^2 w = 0 \end{cases} \tag{4.10}$$

The solution of the above set of equation are given by,

$$\begin{cases} u(x) = A_1\, e^{-ik_x x} + A_2\, e^{ik_x x} \\ v(y) = B_1\, e^{-ik_y y} + B_2\, e^{ik_y y} \\ w(x) = C_1\, e^{-ik_z z} + C_2\, e^{ik_z z} \end{cases} \tag{4.11}$$

Alternately, the solution can be expressed in trignometric functions i.e,

$$\begin{cases} u(x) = A_1 \cos k_x x + A_2 \sin k_x x \\ v(y) = B_1 \cos k_y y + B_2 \sin k_y y \\ w(x) = C_1 \cos k_z z + C_2 \sin k_z z \end{cases} \tag{4.12}$$

So, the complete solution is the product of u, v and w i.e,

$$E_x(x, y, z) = u(x)\, v(y)\, w(z) \tag{4.13}$$

Note that the solution expressed in exponential form represents traveling/propagating waves in all the direction, whereas, trignometric form expresses standing waves in all the directions. Based on the boundary condition, the solutions are chosen and used.

4.2 Polarization of Light and TEM Waves

By definition, the polarization of light is the time course of the electric field vector, $\vec{E}(\vec{r}, t)$ as it propagates (see, Fig. 4.1). To understand, consider a monochromatic plane wave of

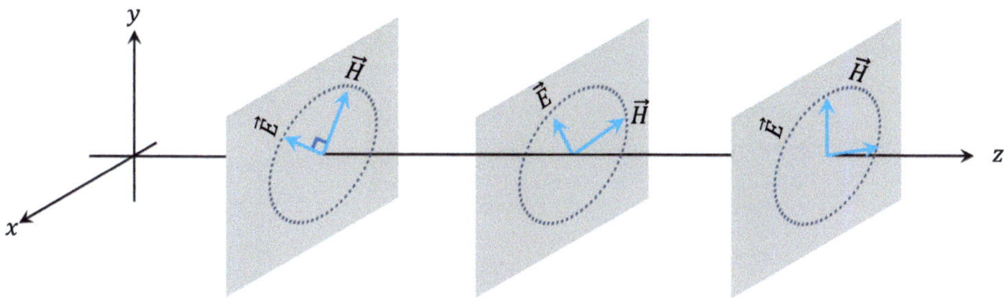

Fig. 4.1 Time course of electric and magnetic field vectors

angular frequency $\omega = 2\pi\nu$ propagating along z,

$$\vec{E}(z,t) = \vec{A}(x,y)\ \cos\left[i\omega\left(t - \frac{z}{c}\right)\right] \tag{4.14}$$

where, $\vec{A} = A_x\hat{i} + A_y\hat{j}$, $j = \sqrt{-1}$, and c is the speed of plane wave.

Since the wave is propagating along z-axis, the complex field components lies in the XY-plane. Representing both the complex components as, $A_x = |A_x|\ e^{i\phi_x}$ and $A_y = |A_y|\ e^{i\phi_y}$, the field can be expanded as,

$$\vec{E}(z,t) = E_x(z,t)\ \hat{i} + E_y(z,t)\ \hat{j} \tag{4.15}$$

with the field components expressed as,

$$\begin{cases} E_x = |A_x|\ \cos\left[i\omega\left(t - \frac{z}{c}\right) + \varphi_x\right] \\ \\ E_y = |A_y|\ \cos\left[i\omega\left(t - \frac{z}{c}\right) + \varphi_y\right] \end{cases} \tag{4.16}$$

Note that, the above set of equations are the parametric forms of an ellipse i.e,

$$\left(\frac{E_x}{|A_x|}\right)^2 + \left(\frac{E_y}{|A_y|}\right)^2 - 2\left(\frac{E_x E_y}{|A_x||A_y|}\right)\ \cos\Delta\varphi - \sin^2\Delta\varphi = 0 \tag{4.17}$$

where, $\Delta\varphi = (\varphi_x - \varphi_y)$ is the phase difference between the components

The above equation (Eq. 4.17) is the master equation describing all possible polarization state of light. The elliptic nature of of the polarization ellipse (Eq. 4.16) states that the tip of electric field rotates helically along z-direction as the wave progresses and repeats its motion with a periodicity of λ. This is pictorially illustrated in Fig. 4.2. Lets us use the master equation (Eq. 4.17), and understand some of the known polarization states of light.

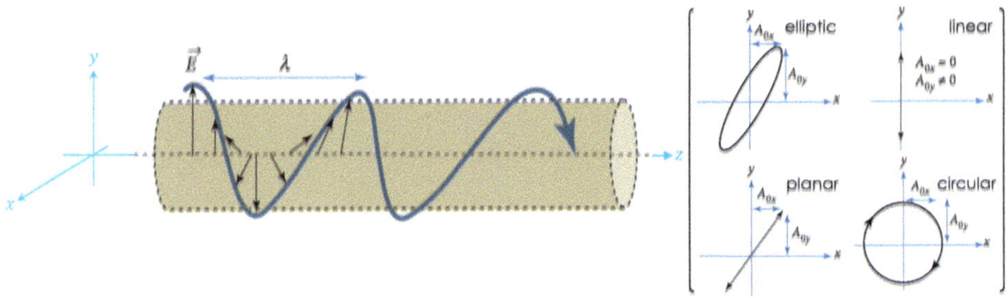

Fig. 4.2 Time course rotation of electric field vectors as the wave propagates. The profile of the electric field in the XY plane as visualized from propagation direction (z-axis). Pictorial representation of four different polarizations (elliptic, linear, planar, circiular). The magnetic field vector is not shown

Linear Polarization If $|A_x| = 0$, then we are left with only y-component, producing linearly-polarized light along y-direction. This is as shown in Fig. 4.2.

Planar Polarization Consider the case, when phase-difference is an even integer multiple of $\pi/2$ i.e, $\Delta\varphi = 0 \; or \; \pi$. The master equation collapses to the following simplified form,

$$\left(\frac{E_x}{|A_x|} + \frac{E_y}{|A_y|}\right)^2 = 0 \quad \Rightarrow E_y = \pm\frac{|A_y|}{|A_x|} E_x \tag{4.18}$$

The above equation represents a straight line with a polarization slope of, $|A_y|/|A_x|$. Assuming, equal magnitude for the components ($|A_x| = |A_y|$), gives, $E_y = E_x$ and the polarization angle is, $\gamma = \tan^{-1}(1) = 45°$ (see, Fig. 4.2).

Circular Polarization Now, consider the case when the phase-difference is odd integer multiple of $\pi/2$. This produces the following polarization components,

$$\begin{cases} E_x = A_0 \cos[i\omega(t - z/c) + \phi_x] \\ E_y = \pm A_0 \sin[i\omega(t - z/c) + \phi_x] \end{cases} \tag{4.19}$$

So, the polarization ellipse simplifies to,

$$E_x^2 + E_y^2 = A_0^2 \tag{4.20}$$

which is a circle with radius, A_0.

The above Eq. 4.20 represents a circularly-polarized light (see, Fig. 4.2). Depending upon the phase relationship, the wave is either right or left circularly polarized. For $\Delta\phi = -\pi/2$ the wave is right-circularly polarized, whereas it is left-circularly polarized for $\Delta\phi = +\pi/2$. Apart from these polarization states, the most general polarization state is represented by Poincare sphere [1].

With the basic understanding of electromagnetic nature, polarization states and diffraction of light, we are set to explore the field distribution at the focus of optical microscopy systems. This is central to optical microscopy and imaging.

4.3 Angular Spectrum Representation

The diffraction of light determines the behavior of light waves in the far-field. In this regard, angular spectrum representation is helpful in determining field distribution far from the source. In optical microscopy, the field distribution is desired in the far-field, specifically at the geometrical focus of the objective lens. Generally, angular spectrum analysis takes into account field propagation and the determines field components.

Consider that the field is known at $z = 0$, and the field is desired at $z = z_0$. In simple terms, the angular spectrum representation connects the space domain and $k -$ domain or frequency space. Mathematically, the angular spectrum representation of the electric field can be expressed as,

$$\vec{E}(x, y, z) = \int\int_{-\infty}^{\infty} \vec{E}'(k_x, k_y, 0)\, e^{i(k_x x + k_y y + k_z z)}\, dk_x\, dk_y \tag{4.21}$$

Here, we are interested in the far-field (say, $r = r_\infty$), so we use asymptotic approximation of the field. To facilitate, we introduce the following unit vectors in the direction $\vec{r}_\infty$,

$$\hat{s} = (s_x, s_y, s_z) = \left(\frac{x}{r}, \frac{y}{r}, \frac{z}{r}\right) \tag{4.22}$$

where, $r = \sqrt{x^2 + y^2 + z^2}$.

Using the above unit vectors, the field can be rewritten as,

$$\vec{E}_\infty(s_x, s_y, s_z) = \lim_{kr\to\infty} \int\int_{(k_x^2+k_y^2)\le k^2} \vec{E}'(k_x, k_y, 0)\, e^{ir(k_x s_x + k_y s_y + k_z s_z)}\, dk_x\, dk_y \tag{4.23}$$

where, $(k_x^2 + k_y^2) > k^2$ represents contribution from near-field i.e, evanescent waves, and we are interested in the far field i.e. $(k_x^2 + k_y^2) \le k^2$.

Next, we use stationary phase to evaluate the double integral. For large kr (i.e, $kr \to \infty$), the following relation holds good [2],

$$-i\frac{2\pi k}{r}\, s_z\, \vec{E}'(-k_x, -k_y)\, e^{-ikr} = \begin{cases} \vec{E}(x, y, z), & p(x, y, z) \in D \\ 0, & p(x, y, z) \notin D \end{cases} \tag{4.24}$$

where, D and $p(x, y, z)$ are the aperture and a point in the exit pupil.

So, the far-field (given by, Eq. 4.23) becomes,

$$\vec{E}_\infty(s_x, s_y, s_z) = -i2\pi k s_z\, \vec{E}'(ks_x, ks_y, 0)\, \frac{e^{ikr}}{r} \tag{4.25}$$

In the above equation, replacing ks_x and ks_y by k_x and k_y immediately gives the far-field as Fourier spectrum, $E'(k_x, k_y, 0)$. Representing the far-field Fourier spectrum by $\vec{E}_\infty(k_x, k_y)$ we can write, $E_\infty(s_x, s_y, s_z) = E_\infty(k_x, k_y)$. This modifies the above relation (Eq. 4.25) to,

$$\vec{E}'(k_x, k_y, 0) = i\frac{r\, e^{-ikr}}{2\pi k_z}\, \vec{E}_\infty(k_x, k_y) \tag{4.26}$$

Now, substituting Eq. 4.26 in Eq. 4.21 we get,

$$\vec{E}(x, y, z) = \frac{ir\, e^{-ikr}}{2\pi k_z} \int\int_{(k_x^2+k_y^2)\leq k^2} \vec{E}_\infty(k_x, k_y)\, e^{i(k_x x + k_y y + k_z z)}\, dk_x\, dk_y \tag{4.27}$$

This is the angular spectrum representation of the optical field at a plane far-off (say, focus) from the source.

4.4 Field Distribution at Geometric Focus

Most of optical microscopy systems are single-lens based i.e, have a single field focussing element. Here, we are interested in the distribution of the electric field at the focus of a lens. Figure 4.3 shows wave propagation in an aplanatic system where the lens is illuminated by an uniform electromagnetic plane wave for which the electric and magnetic field vectors

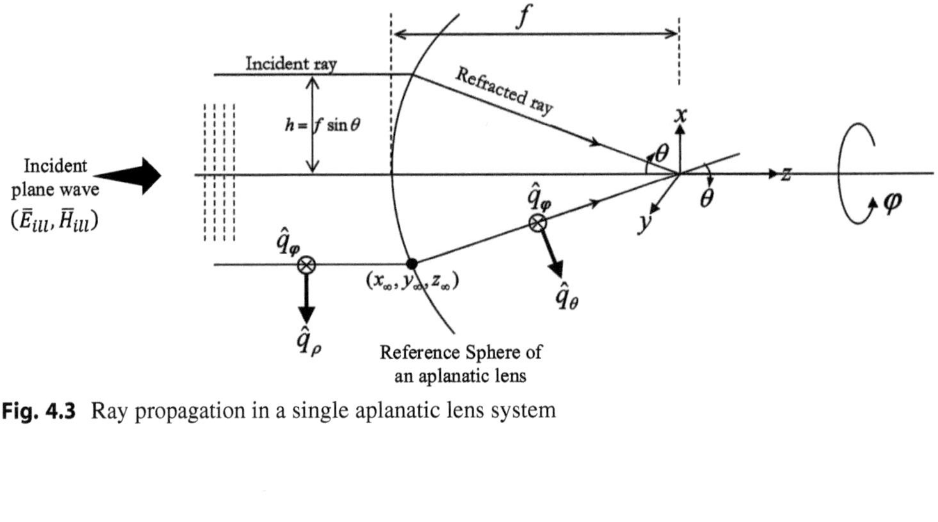

Fig. 4.3 Ray propagation in a single aplanatic lens system

are given by,

$$\vec{E}_{ill} = re\{\vec{E}_{ill}\, e^{-i\omega t}\}\,, \quad \vec{H}_{ill} = re\{\vec{H}_{ill}\, e^{-i\omega t}\} \tag{4.28}$$

The lens system transforms an incoming plane wavefront into a spherical wavefront (see, Fig. 4.3). Consider that, $(\hat{q}_\rho, \hat{q}_\varphi)$ and $(\hat{q}_\theta, \hat{q}_\varphi)$ as the unit vectors in the cylindrical and spherical coordinate system, respectively. Accordingly, the electric field can be represented by two orthogonal components, $\vec{E}^s$ and $\vec{E}^p$,

$$\vec{E}^s = (\vec{E}_{ill}.\hat{q}_\varphi)\,\hat{q}_\varphi\,, \quad \vec{E}^p = (\vec{E}_{ill}.\hat{q}_\rho)\,\hat{q}_\rho \tag{4.29}$$

where, the indices s and p represent s and p polarizations, respectively.

As the wave crosses the air-glass interface created by the first lens surface, the vector $\hat{q}_\rho$ gets mapped to $\hat{q}_\theta$ while the vector $\hat{q}_\phi$ remains unchanged. Using intensity law,[3] the refracted electric field $\vec{E}_\infty$ can be expressed as,

$$\begin{aligned}\vec{E}_\infty &= \sqrt{\frac{n_1}{n_2}}\,[T^s\vec{E}^s + T^p\vec{E}^p]\,\cos^{1/2}\theta \\ &= \sqrt{\frac{n_1}{n_2}}\,[T^s(\vec{E}_{ill}.\hat{q}_\varphi)\,\hat{q}_\varphi + T^p\,[(\vec{E}_{ill}.\hat{q}_\rho)\,\hat{q}_\theta]\,\cos^{1/2}\theta\end{aligned} \tag{4.30}$$

where, T^s and T^p are the transmission coefficients.

The next step is to express the unit vectors, $(\hat{q}_\rho, \hat{q}_\varphi, \hat{q}_\theta)$ in Cartesian coordinates, $(\hat{q}_x, \hat{q}_y, \hat{q}_z)$ using the following transformation matrix,

$$M_T = \begin{bmatrix} \cos\varphi & \sin\varphi & 0 \\ -\sin\varphi & \cos\varphi & 0 \\ \cos\theta\cos\varphi & \cos\theta\sin\varphi & \sin\theta \end{bmatrix} \tag{4.31}$$

and the complete relation is given by,

$$\begin{bmatrix} \hat{q}_\rho \\ \hat{q}_\varphi \\ \hat{q}_\theta \end{bmatrix} - M_T \begin{bmatrix} \hat{q}_x \\ \hat{q}_y \\ \hat{q}_z \end{bmatrix} \tag{4.32}$$

[3] Intensity law: The field before and after refraction are related by, $|\vec{E}_2| = \sqrt{\frac{n_1}{n_2}}\,|\vec{E}_1|\,cos^{1/2}\theta$

Expansion produces the following set of equations,

$$\begin{bmatrix} \hat{q}_\rho = \cos\varphi\ \hat{q}_x + \sin\varphi\ \hat{q}_y \\ \hat{q}_\varphi = -\sin\varphi\ \hat{q}_x + \cos\varphi\ \hat{q}_y \\ \hat{q}_\theta = \cos\theta\ \cos\varphi\ \hat{q}_x + \cos\theta\ \sin\varphi\ \hat{q}_x + \sin\theta\ \hat{q}_z \end{bmatrix} \tag{4.33}$$

Now going back to our problem as shown in Fig. 4.3, we consider the incident beam to be polarized along x-axis (s-polarized) with high transmissivity ($T^s = 1$), i.e, $\vec{E}_{ill} = E_{ill}\ \hat{q}_x$. From Fig. 4.3, q_x have two components i.e., $\hat{q}_\varphi$ and $\hat{q}_\theta$ in the spherical coordinate. Accordingly, we have the following expression for the refracted field,

$$\vec{E}_\infty(\theta,\phi) = E_{ill}(\theta,\varphi)\ \underbrace{[\cos\varphi\ \hat{q}_\theta - \sin\varphi\ \hat{q}_\varphi]}_{\hat{q}_x}\ \sqrt{n_1/n_2}\ \cos^{1/2}\theta \tag{4.34}$$

Using Eq. 4.33 and expansion gives the following form,

$$\vec{E}_\infty(\theta,\phi) = \frac{1}{2}\ E_{ill}(\theta,\varphi) \begin{bmatrix} (1+\cos\theta) - (1-\cos\theta)\ \cos 2\varphi \\ -(1-\cos\theta)\ \sin 2\varphi \\ 2\sin\theta\ \cos\varphi \end{bmatrix} \sqrt{n_1/n_2}\ \cos^{1/2}\theta \tag{4.35}$$

We use angular spectrum representation in k-space to relate the field at the focus, $\vec{E}(\rho,\phi,z)$ and the refracted field $\vec{E}_\infty(\theta,\varphi)$,

$$\vec{E}(\rho,\phi,z) = i\frac{kf\ e^{ikf}}{2\pi} \int_{\theta=0}^{\alpha}\int_{\varphi=0}^{2\pi} \vec{E}_\infty(\theta,\varphi)\ e^{ik\rho\sin\theta\cos(\varphi-\phi)}\ e^{ikz\cos\theta}\ \sin\theta\ d\theta\ d\varphi \tag{4.36}$$

where we have use the following transformations,

$$\begin{cases} k_x = k\sin\theta\ \cos\varphi \\ k_y = k\sin\theta\ \sin\varphi \\ k_z = k\ \cos\theta \end{cases} \quad \begin{cases} x = \rho\cos\theta \\ y = \rho\sin\theta \\ \frac{1}{k_z}dk_x dk_y = k\sin\theta\ d\theta\ d\varphi \end{cases} \tag{4.37}$$

Considering uniform plane wave illumination i.e, $E_{ill}(\theta,\varphi) = E_0$, and substituting Eq. 4.35 in Eq. 4.36 produces the explicit expression for the field at the focus,

$$\vec{E}(\rho,\phi,z) = i\frac{kf\ e^{ikf}}{4\pi} \int_{\theta=0}^{\alpha}\int_{\varphi=0}^{2\pi} E_0 \begin{bmatrix} (1+\cos\theta) - (1-\cos\theta)\ \cos 2\varphi \\ -(1-\cos\theta)\ \sin 2\varphi \\ 2\sin\theta\ \cos\varphi \end{bmatrix}$$

$$\times\ e^{ik\rho\sin\theta\cos(\varphi-\phi)}\ e^{ikz\cos\theta}\ \sin\theta\ \cos^{1/2}\theta\ d\theta\ d\varphi \tag{4.38}$$

Next, we simplify the above expression by incorporating all the constants in E_0, and use the following well-known mathematical identities involving Bessel function,

$$\begin{cases} \int_0^{2\pi} \cos(n\varphi)\, e^{ix\cos(\varphi-\phi)}\, d\varphi = i^n\, 2\pi\, J_n(x)\, \cos(n\phi) \\ \int_0^{2\pi} \sin(n\varphi)\, e^{ix\cos(\varphi-\phi)}\, d\varphi = i^n\, 2\pi\, J_n(x)\, \sin(n\phi) \end{cases} \tag{4.39}$$

Using the above identities, Eq. 4.38 becomes,

$$\vec{E}(\rho, \phi, z) = E_0 \begin{bmatrix} I_0 + I_2\, \cos 2\phi \\ I_2\, \sin 2\phi \\ -i2I_1\, \cos\phi \end{bmatrix} \tag{4.40}$$

where the diffraction integrals are given by,

$$\begin{cases} I_0 = \int_0^{\alpha} \cos^{1/2}\theta\, \sin\theta(1+\cos\theta)\, J_0(k\rho\sin\theta)\, e^{ikz\cos\theta}\, d\theta \\ I_1 = \int_0^{\alpha} \cos^{1/2}\theta\, \sin^2\theta\, J_1(k\rho\sin\theta)\, e^{ikz\cos\theta}\, d\theta \\ I_2 = \int_0^{\alpha} \cos^{1/2}\theta\, \sin\theta\, (1-\cos\theta)\, J_2(k\rho\sin\theta)\, e^{ikz\cos\theta}\, d\theta \end{cases} \tag{4.41}$$

Equation 4.40 is the integral representation of the field at and near the focus of an aplanatic lens.

The intensity distribution at and near the focus can be expressed as,

$$I(\rho, \phi, z) = |\vec{E}(\rho, \phi, z)|^2 = E_x^2 + E_y^2 + E_z^2 \tag{4.42}$$

where, the field components are given by Eq. 4.40.

The above set of equation can be computed for visualizing field distribution at and near the geometrical focus of an aplanatic lens. This is really beneficial since this allows field determination purely based on computer simulation and does not involve actual field measurements. Equation 4.40 is very helpful to analyze a variety of imaging systems as we will discuss in the subsequent sections.

4.5 Fluorescence Microscopy Systems

In this section we will be discussing modern fluorescence based microscopy and imaging systems. These systems forms the basis of super-resolution microscopy. In general, optical microscopy systems are better characterized by their point spread function (PSF).

Sometimes they are also refereed to as system PSF. We will be using both terminology in this textbook.

The system PSF of a microscopy system is essentially the product of excitation and detection PSFs in the space domain i.e,

$$h_{sys}(x, y) = h_{exc}(x, y) \times h_{det}(x, y) \tag{4.43}$$

Note that, the Fourier transform of the system PSF i.e., H_{sys} is popularly termed as optical transfer function (OTF). Since the point source undergoes diffraction during propagation through focusing element and free-space (lens to focal plane), it is the image of the point which appears as an airy-disc pattern. This is also known as *point spread function.*

4.5.1 Widefield Microscopy

This is the simplest and most widely used fluorescence-based microscopy technique. The schematic diagram of a typical widefield fluorescence system with an inverted optical configuration is shown in Fig. 4.4. In this configuration, light from a laser of appropriate wavelength is used to illuminate the fluorescent specimen. It may be noted that the choice of laser of specific wavelength depends on the fluorophore that is used to label the specimen. The light is appropriately expanded with the help of a beam-expander (BE) that comprises two lenses of different focal lengths (say, f_1 and f_2 with $f_2 > f_1$), for which the expansion factor is $M = f_2/f_1$. This factor is chosen so that the beam slightly overfills the back aperture of the objective lens, which helps in availing the complete aperture of the lens. Upon expansion, the beam is then directed to the objective lens using a dichroic mirror that allows the incident wavelength (say, λ_{ill}) to reflect, but pass through wavelength $> \lambda_{ill}$. Overfilled back-aperture ensures the use of maximum numerical aperture (NA) of the objective lens, thereby tightly focusing the light, resulting in a point-like concentrated beam at the specimen (see, Fig. 4.4). Filters (F_1 and F_2) may

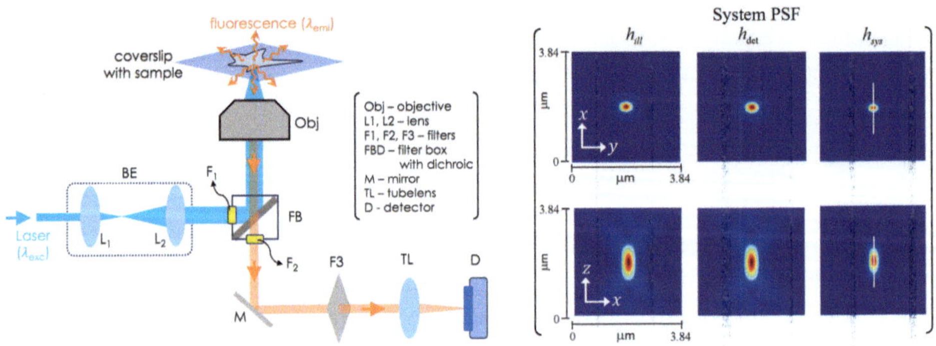

Fig. 4.4 Schematic of a typical widefield fluorescence microscopy. Alongside key optical components are also listed. The excitation, emission and system PSF is shown on the right half

be used to eliminate the background light. The excitation PSF or the field distribution at the focus can be computationally determined from the relations (4.40 and 4.42) derived in the previous section. Accordingly, the excitation PSF (for randomly-polarized light) is given by,

$$h_{exc} = |E_{ill}(x, y)|^2 = |E_x|^2 + |E_y|^2 + |E_z|^2$$

$$= |I_{0,\lambda_{ill}}|^2 + 2\,|I_{1,\lambda_{ill}}|^2 + |I_{2,\lambda_{ill}}|^2 \tag{4.44}$$

where the diffraction integrals are calculated for a wavelength of λ_{ill} using Eq. 4.41.

Upon fluorescence emission from the specimen, a large portion (equivalent to the solid angle represented by α) of it is collected by the objective lens and passed on to the detection module.[4] A portion of the fluorescence light is collected by the objective lens, and a parallel beam of fluorescence light passes through the dichroic mirror (DM), which is followed by wavelength filtering by an additional filter (F_3). The filtered light is then focused by the tube lens (TL) onto the detector, where a magnified image of the object is formed. Here, the detection PSF can be computationally determined using Eqs. 4.40 and 4.42 for a stoke-shifted wavelength (λ_{det}). Accordingly, the detection PSF is given by,

$$h_{det} = |E_{det}(x, y)|^2 = |I_{0,\lambda_{det}}|^2 + 2\,|I_{1,\lambda_{det}}|^2 + |I_{2,\lambda_{det}}|^2 \tag{4.45}$$

It may be noted that, the above relations (Eqs. 4.44 and 4.45) are derived for the simplest possible fluorescence microscope. A real microscope may be equipped with additional features (filters, mirrors, lenses) to enhance image quality (magnification and SBR). Computational simulation on a 256 × 256 pixel grid generates the excitation, emission and system PSF as displayed in Fig. 4.4.

4.5.2 Confocal Fluorescence Microscopy

The idea of confocal microscopy was first put forward by Minsky[5] [3]. This has ushered the way to many modern fluorescence microscopes. Repercussions can be seen in a number of fields ranging from material science to clinical diagnostics. Now a days, confocal microscopy is extensively used for in-vivo imaging of tissues and organs. Recently, it is used to access drug delivery and efficacy studies. Confocal is a valuable tool for accessing the progression of various diseases including, cancer and neurodegenerative disorders. Of late it is used in clinical fields such as, opthalmology and dermatology. Moreover,

[4] Noted that fluorescence emission is isotropic in nature, i.e., emission occurs in complete solid 4π angle.

[5] M. Minsky Microscopy apparatus, US Patent 3013467, Dec 1961.

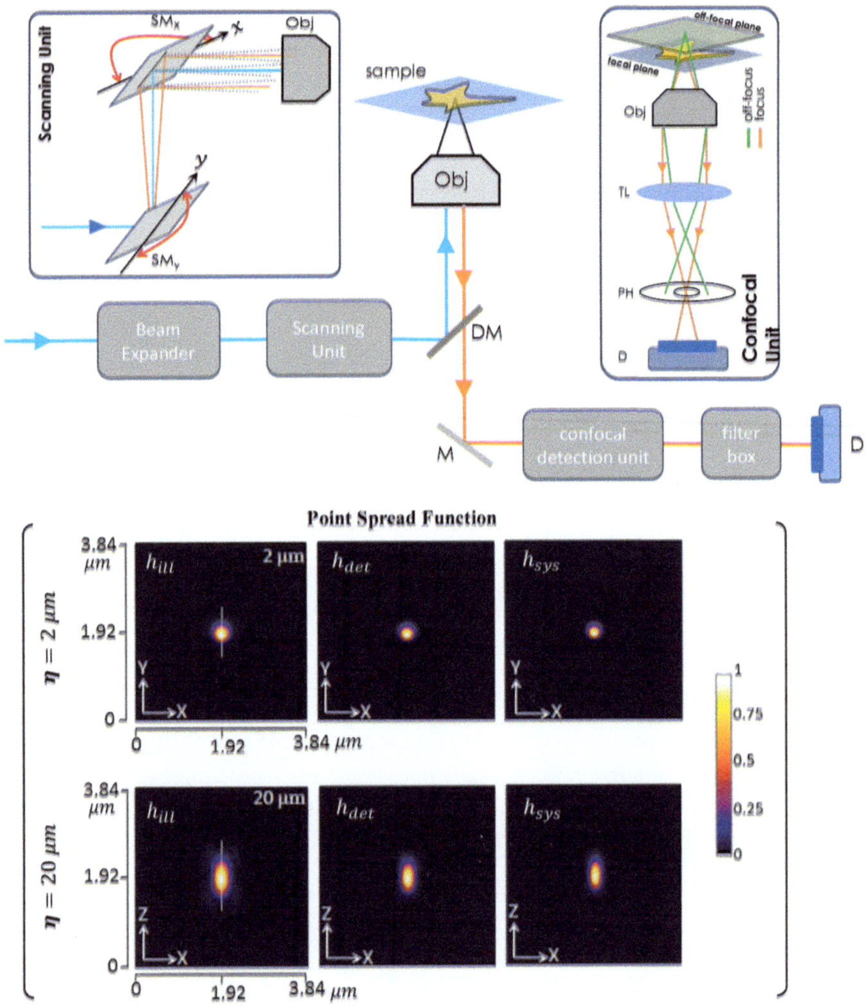

Fig. 4.5 Schematic of a typical scanning confocal microscopy system. The inset shows the scanning and confocal units. Alondside, the excitation, emission and system PSFs are also shown

confocal is a key technique for determining physical state of the matter at microscopic scales. Many more fields are embracing confocal microscopy for detailed visualization of the fine structures and access biological processes under investigation.

Although widefield microscopy produces microscopic images of the specimen, it requires further processing to get rid of blur which is predominantly due to out of focus light reaching the detector. Conceptually this can be tackled by eliminating light coming from off-focal planes as done in confocal microscope. A schematic diagram of a typical confocal system is shown in Fig. 4.5. Apart from the optics used in widefield microscopy, two additional optical units are used in confocal scanning microscopy: point scanning unit and confocal unit. The point scanning unit has a X-scanning mirror (SM_X) and a Y-scanning mirror (SM_Y) which work in synchrony while scanning a region/area of the sample. The beam moving laterally (along X and Y) at the back-aperture of the objective lens, causes the focus to scan over the specimen. In the detection sub-system, the fluorescence from the specimen (originating from all the off-focal and focal planes) is collected by the same objective lens and diverted to the confocal unit. The inset in Fig. 4.5 shows the confocal unit. Light from the focal plane gives rise to parallel beams which upon focusing by the tube lens (TL) passes through a pin-hole (placed just before the detector). However, light from off-focal planes gives rise to non-parallel rays (both convergent and divergent) which upon focused by the tube lens (TL) gets blocked by the pin-hole (PH). So, the pin-hole efficiently eliminates the light from off-focal planes and allows only light from focal plane to reach the detector. However, the implementation of the unit may somewhat differ from the optical configuration shown in Fig. 4.5. The detector normally used is a point detector such as photomultiplier tube. The fluorescence/signal is collected in a point-by-point manner until the entire chosen area of the sample is scanned. This gives rise to a crisp blur-free image of the specimen. Finally, the resulting image is a collage of all the points scanned in the object plane/specimen.

For a confocal system, the illumination PSF is similar to that of widefield (apart from the scanning unit),

$$h_{exc} = |I_{0,\lambda_{ill}}|^2 + 2\,|I_{1,\lambda_{ill}}|^2 + |I_{2,\lambda_{ill}}|^2 \tag{4.46}$$

However, the detection in confocal involves an aperture (pinhole) in front of the detector. In general, a point source does not appear as a point at the image plane (focus), rather gets distorted by diffraction and light originating from above and below the focal plane (off-focal planes). This process is called convolution and is inherent to optical systems. Here the induction of pinhole modifies the wave-propagation by blocking the out-of-focus light thereby convolving with the emission PSF to produce the pinhole-convolved PSF. So, the detection PSF for confocal microscopy is the convolution of widefield detection PSF with the pin-hole function, $\kappa(x, y)$ i.e,

$$h_{det} = [|I_{0,\lambda_{det}}|^2 + 2\,|I_{1,\lambda_{det}}|^2 + |I_{2,\lambda_{det}}|^2] \otimes \kappa(x, y) \tag{4.47}$$

where, κ is the geometry of the pinhole and the diffraction integrals are calculated for the emission wavelength, λ_{emi}.

Computational study can be carried out for an ideal confocal system and the same is shown in Fig. 4.5. Specifically, two different pinholes sizes are considered. The changes are apparent, with small pinhole producing a near symmetrical PSF. However, a small pinhole substantially reduces the light reaching to the point detector. This manifests the important role played by pinhole in confocal microscopy.

4.5.3 Two Photon Microscopy

Most of the fluorescence microscopy techniques require excitation of fluorophores that label specific organelle in the specimen. In general, Jablonski energy diagram is used to understand the excitation of fluorophore upon exposure to a light of specific wavelength (say, λ_{ill}). The excitation process is mostly linear with respect to the illumination intensity. This is what happens in single photon excitation microscopy. However, non-linear processes do exist that may lead to excitation.

Unlike single photon excitation, two photon excitation requires two photons of half the energy compared to that of a single photon interacting with the fluorophore [4–6]. A Jablonski energy diagram describing two-photon excitation process is shown in Fig. 4.6a. In a typical two photon excitation process, two low energy photons almost simultaneously interact with the fluorophore to cause a higher-energy electronic transition (see, Fig. 4.6a). As usual, the electron undergoes internal conversion (taking $\sim 10^{-12}$ sec) to the lowest vibrational level of the first excited state, S_1. Subsequently, the fluorophore relaxes to the ground singlet state (S_0), emitting a Stoke-shifted photon (fluorescence). However, 2PE is a nonlinear process for which the absorption rate depends on the second power of incident light, resulting in fluorescence emission only from a confined region/PSF. This is due to the fact that the intensity is the highest in the vicinity of the focus of a focused light and falls-off instantaneously/quadratically with the distance above and below the ideal focal plane (see, Fig. 4.6b). This increases the probability of two photons simultaneously (in less than femtoseconds) interacting with the molecule, thereby increasing the chance of two-photon excitation at the focal spot. On the contrary, the fluorescence for 1PE is largely from the X-pattern (comprising several planes of the specimen) at the focal volume that

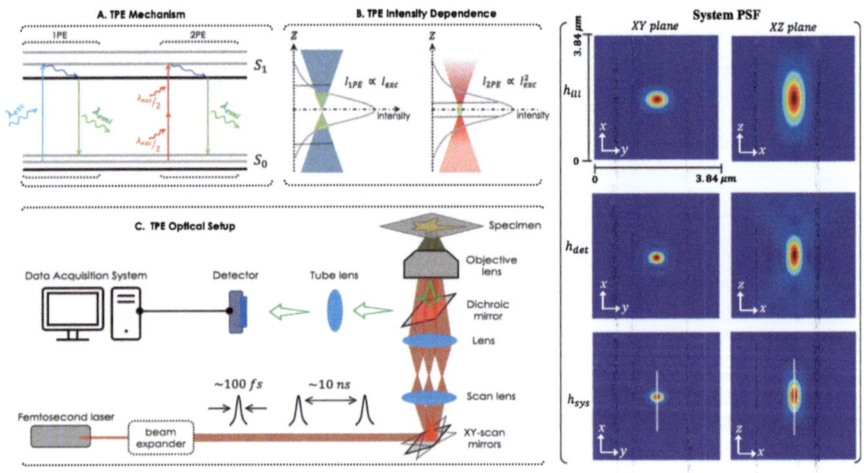

Fig. 4.6 Two-photon excitation microscopy: (**a**) TPE excitation mechanism, (**b**) Two photon intensity dependence, (**c**) TPE optical setup. The excitation, emission and system PSFs are shown on the right side

include fluorescence emission from off-focal planes as well. This is pictorially illustrated in Fig. 4.6b.

So, for 2PE, the excitation PSF is given by,

$$h_{exc}^{2PE} = |I_{exc}|^2 = [\ |I_{0,\lambda_{ill}}|^2 + 2\ |I_{1,\lambda_{ill}}|^2 + |I_{2,\lambda_{ill}}|^2\ ]^2 \tag{4.48}$$

Since the fluorescence emission is isotropic and randomly-polarized and the detection is performed in a widefield mode, the detection PSF is,

$$h_{det} = |I_{0,\lambda_{det}}|^2 + 2\ |I_{1,\lambda_{det}}|^2 + |I_{2,\lambda_{det}}|^2 \tag{4.49}$$

The quadratic dependence and the localization of 2PE effect can be evaluated computational by implementing Eqs. 4.48 and 4.49. Figure 4.6 shows the excitation, detection and system PSF for a typical 2PE microscope.

4.6 Field Dipole Interaction in Fluorescence Microscopy

The interaction of light field with fluorophore dipole is at the heart of a all fluorescence-based optical microscopy systems. Until now, we investigated microscopy techniques where randomly polarized incident field interacts with randomly oriented dipoles. In this section, we will explore interaction of polarized light field with dipoles oriented in specific directions. Such an interaction is dominant when both electric field vector and dipole are perfectly aligned, and are key to understand polarization effects in optical microscopy.

Consider optical setup of a general optical microscope as shown in Fig. 4.7. Upon excitation of the fluorophore dipole, the emission from the object plane O is collected by the objective lens (L_2) and image is formed by the tube lens (L_3) in the detector plane (I). A harmonically-oscillating electric dipole (located at the object plane O) gives rise to a far-field [7],

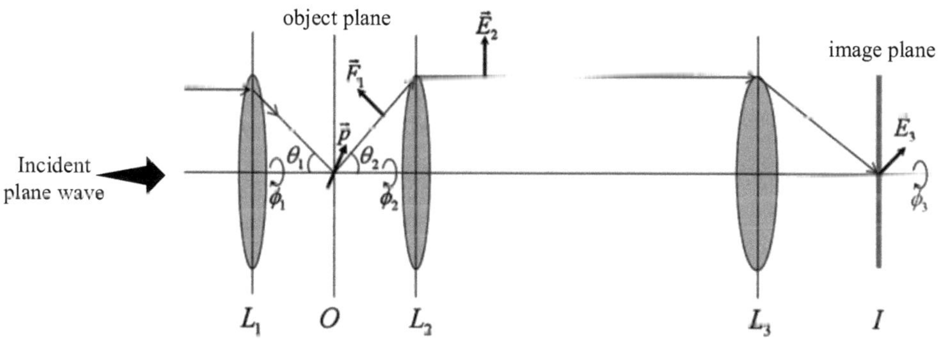

Fig. 4.7 A typical optical microscope showing field-dipole interaction

$$\vec{E}_O = \vec{r} \times (\vec{r} \times \vec{p}) \tag{4.50}$$

where, the constants are ignored for simplicity, and $\vec{r}$ is the position vector.

The light is collected by the objective lens (L_2) and the refracted field can be expressed as,

$$\vec{E}_R = (\vec{p}.\hat{a}_\phi)\,\hat{a}_\phi + [(\vec{r} \times \vec{p}).\hat{a}_\phi]\,\hat{a}_\rho \tag{4.51}$$

where, $\hat{a}_\phi$, $\hat{a}_\rho$, and $\hat{a}_z$ are unit vectors in cylindrical coordinate system.

Note that, $\vec{r} = x\hat{i}+y\hat{j}+z\hat{k} = r\cos\theta\hat{i}+r\sin\theta\,\hat{j}+z\hat{k}$. And the unit vectors in cylindrical coordinate are,

$$\begin{cases} \hat{a}_\rho = \frac{\partial \vec{r}}{\partial r}/|\frac{\partial \vec{r}}{\partial r}| = \cos\theta\;\hat{i} + \sin\theta\;\hat{j} \\ \hat{a}_\phi = \frac{\partial \vec{r}}{\partial \theta}/|\frac{\partial \vec{r}}{\partial \theta}| = -\sin\theta\;\hat{i} + \cos\theta\;\hat{j} \\ \hat{a}_z = \frac{\partial \vec{r}}{\partial z}/|\frac{\partial \vec{r}}{\partial z}| = \hat{k} \end{cases} \tag{4.52}$$

The dipole moments (in spherical coordinate system) can be expressed as,

$$\begin{aligned} \vec{p} &= p_x\;\hat{i} + p_y\;\hat{j} + p_z\;\hat{k} \\ &= \sin\theta\cos\phi\;\hat{i} + \sin\theta\sin\phi\;\hat{j} + \cos\theta\;\hat{k} \end{aligned} \tag{4.53}$$

Substituting the explicit expression for $\vec{p}$ (Eq. 4.53) and unit vectors 4.52 in Eq. 4.51, produces the following expression for the refracted field,

$$\begin{aligned} \vec{E}_2 = {} & [p_x\;(\sin^2\theta + \cos\theta_2\cos^2\phi_2) - p_y\;\sin\phi_2\cos\phi_2(1-\cos\phi_2) - p_z\;\sin\theta_2\cos\phi_2]\;\hat{i} \\ & -[p_x\;\sin\phi_2\cos\phi_2(1-\cos\theta_2) - p_y\;(\cos^2\theta_2 + \cos\theta_2\sin^2\phi_2) - p_z\;\sin\theta_2\sin\phi_2]\;\hat{j} \end{aligned} \tag{4.54}$$

Using angular spectrum representation in a polar coordinate system (Eqs. 4.26 and 4.36), the field in the image plane is given by,

$$\vec{E}_3(\rho_3,\phi_3) = \int_0^{2\pi}\int_0^{\alpha} \vec{E}_2(\theta_2,\phi_2)\,e^{ik\rho_3\sin\theta_2\cos(\phi_2-\phi_3)}\,e^{ikz\cos\theta_2}\;\sin\theta_2\;\cos^{1/2}\theta_2\,d\theta_2\,d\phi_2 \tag{4.55}$$

Finally, substituting $\vec{E}_2$ (Eq. 4.54) in $\vec{E}_3$ (Eq. 4.55), and further simplification yields the following general expression,

$$\begin{aligned} I = |\vec{E}_3|^2 = {} & (p_x^2 + p_y^2)\;[|I_0|^2 + |I_2|^2] + 2[(p_x^2 - p_y^2)\;\cos 2\phi + 2p_x p_y\sin 2\phi]\;re\{I_0 I_2^*\} \\ & +4\;p_z^2\;|I_1|^2 \end{aligned} \tag{4.56}$$

where the diffraction integrals are,

$$\begin{cases} I_0 = \int_0^{\alpha} \sin\theta(1+\cos\theta)\ J_0(v'/\sin\alpha)\ e^{iu'/sin^2\alpha}\ \cos^{1/2}\theta\ d\theta \\ I_1 = \int_0^{\alpha} \sin^{\theta}\ J_1(v'/\sin\alpha)\ e^{iu'/sin^2\alpha}\ \cos^{1/2}\theta\ d\theta \\ I_2 = \int_0^{\alpha} \sin\theta(1-\cos\theta)\ J_2(v'/\sin\alpha)\ e^{iu'/sin^2\alpha}\ \cos^{1/2}\theta\ d\theta \end{cases} \tag{4.57}$$

and, $v' = v\sin\theta$, $u' = u\cos\theta$.

The above set of equations forms the basis to determine field distribution in the image plane, considering field-dipole interaction. In the next subsection, we will use these equations to understand the effect of polarization and compute field distribution for epifluorescence, confocal and scanning microscopy.

4.6.1 Epifluorescence Microscopy

Consider the case of epifluorescence microscopy with freely-rotating dipole interacting with the incident field. In this case, the fluorescent molecule/dipole can rotate between the excitation and emission.

Now, substituting explicit expression for dipole moment components (Eq. 4.53) in the intensity expression (Eq. 4.56) gives,

$$I = [|I_0|^2 + |I_2|^2 + 2\,re\{I_0 I_2^*\}\ \cos 2(\phi_p - \phi)]\ \sin^2\theta + 4|I_1|^2\ \cos^2\theta \tag{4.58}$$

Since freely-rotating dipole average over all orientations of dipole, the resultant intensity at the focal plane is given by,

$$\begin{aligned} I_{avg} &= \int_0^{\pi}\int_0^{2\pi} I\sin\phi_p\ d\phi_p\ d\theta_p \\ &= \frac{8\pi}{3}\ [|I_0|^2 + 2|I_1|^2 + |I_2|^2] \end{aligned}$$

4.6.2 Scanning Fluorescence Microscopy

In this type of microscopy, two specific cases arises: first, polarized light that excites dipole in the direction of the electric field. Second, randomly-polarized light interacting with fluorophores/dipoles.

Linearly Polarized Light Illumination The illumination/excitation light is polarized in the x-direction. Following Eq. 4.40, the field at the geometric focus is given by,

$$\vec{E}(\rho, \phi, z) = E_0 \begin{bmatrix} I_0 + I_2 \cos 2\phi \\ I_2 \sin 2\phi \\ -j2I_1 \cos\phi \end{bmatrix} \tag{4.59}$$

where, the constants are ignored.

In general, the image intensity can be expressed as,

$$I = |\vec{p} \cdot \vec{E}|^2 = |p_x E_x + p_y E_y + p_z E_z|^2 \tag{4.60}$$

However, the intensity for dipole oriented along x-direction is,

$$\begin{aligned} I_{p_x} = |p_x E_x|^2 &= C\, |I_0 + I_2 \cos 2\phi|^2 \\ &= |I_0|^2 + |I_2|^2 \cos^2 2\phi + 2\{I_0 I_2^*\} \cos 2\phi \end{aligned} \tag{4.61}$$

where, $|p_x| = C$ is a constant.

Similarly, the intensity for dipoles oriented along y and z direction are,

$$\begin{cases} I_{p_y} = |p_y \cdot E_y|^2 = |I_2|^2 \sin^2 2\phi \\ I_{p_z} = |p_z \cdot E_z|^2 = 4|I_1|^2 \cos^2\phi \end{cases} \tag{4.62}$$

However, in the case of all possible dipole orientations and x-polarized light illumination, the image intensity is obtained by averaging over all dipole orientation i.e,

$$\begin{aligned} I_{x-pol} &= \int_0^{2\pi} (I_{p_x} + I_{p_y} + I_{p_z})\, d\phi \\ &= |I_0|^2 + |I_2|^2 + 2re\{I_0 I_2^*\} \cos 2\phi + 4|I_1|^2 \cos^2\phi \end{aligned} \tag{4.63}$$

Randomly Polarized Light Illumination The incident light is assumed randomly-polarized for which the image intensity is given by Eq. 4.56 i.e,

$$\begin{aligned} I = (p_x^2 + p_y^2)\, [|I_0|^2 + |I_2|^2] + 2[(p_x^2 - p_y^2) \cos 2\phi + 2p_x p_y \sin 2\phi]\, re\{I_0 I_2^*\} \\ + 4\, p_z^2\, |I_1|^2 \end{aligned} \tag{4.64}$$

The image intensity for x-oriented dipoles is obtained by substituting $p_y = p_z = 0$ i.e,

$$I_{p_x} = |I_0|^2 + |I_2|^2 + 2\, re\{I_0 I_2^*\} \cos 2\phi \tag{4.65}$$

Similarly for y and z oriented dipoles, the image intensity is given by,

$$\begin{cases} I_{p_y} = |I_0|^2 + |I_2|^2 - 2\,\{I_0 I_2^*\}\,\cos 2\phi \\ I_{p_z} = 4\,|I_1|^2 \end{cases} \tag{4.66}$$

All these cases for different microscopy techniques suggests that the effect of field polarization and the orientation of fluorophore dipole has a critical impact on the detected image. A detailed description of polarization effects for other microscopy techniques can be found in Ref. [7]. So, precision microscopy requires consideration of polarization of light and field-dipole interaction.

References

1. Saleh BEA, Teich MC (2012) Fundamentals of photonics, 2nd edn. Wiley, New York
2. Wolf E (1959) Electromagnetic diffraction in optical systems. In: An integral representation of the image field. Proceedings of the Royal Society of London. Series A, vol 253, pp 349–357
3. Misky M, Misky M (1988) Memoir on inventing the confocal scanning microscope. Scanning 10:128–138
4. Goeppert-Mayer M (1931) Über Elementarakte mit zwei Quantensprüngen. Annals of Physics 9(3):273–295
5. Göppert-Mayer M (2009) Elementary processes with two quantum transitions. Annalen der Physik 18(7–8):466–479
6. Denk W, Stricker JH, Webb WW (1990) Two-photon laser scanning fluorescence microscopy, Science 248:4951
7. Sheppard CJR, Torok P (1999) An electromagnetic theory of imaging in fluorescence microscopy and imaging in polarization fluorescence microscopy. Bioimaging 5:205–218

Quantum Optics for Precision Microscopy

5

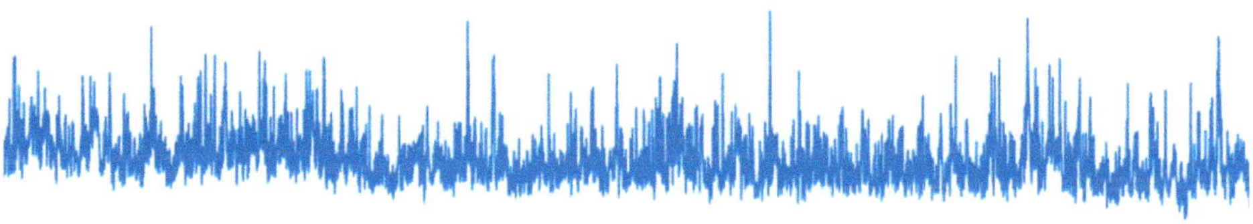

The fact that light exists in minuscule chunks of energy packets ($\sim \hbar\nu$) gives it a distinct character. It appears that light can exist or be prepared in a number of quantum states: Fock state, coherent state, squeezed state and numerous other states. In fact, there are instances where light can exist in a mixed state, i.e., a classical state with local quantum behavior. These states with bizarre quantum behavior have unique properties. Specifically, light in a specific quantum state (quantum entangled state or squeezed state) gives rise to precision measurement of quantities (phase or intensity) beyond the classical limit. In addition, entanglement can be exploited to improve contrast, resolution, and sensitivity. These peculiar properties have repercussions in precision optical microscopy. The quantum state of light is known to play a critical role in super-resolving features. In this chapter, we will discuss all these aspects, beginning with some of the basic quantum states of light and exploring its potential in optical microscopy.

The arrival of quantum optics in optical microscopy has generated new interest. This is largely due to the awkward nature of light when it is in a quantum state. In this state, light behaves more as a particle than a wave, giving it unique properties that are not apparent in classical light. These unique properties can be exploited for precision measurements and imaging in optical microscopy.

Light, as we know, largely exhibits wave nature that forms the basis for classical microscopy. This includes diffraction which ultimately caps the limit on optical resolution. Of late, a few classical techniques (STED, SMLM, SIM, and others) have been demon-

P. P. Mondal, S. Hess, *Single-Molecule Super-Resolution Optical Microscopy*,
https://doi.org/10.1007/978-3-032-13442-4_5

strated to overcome this limit. The next big step in microscopy could be the exploitation of quantum optical effects in super-resolution microscopy to push the limit of intensity or phase measurements. Some of the quantum optical techniques that can be used are quantum entanglement and squeezed light. Special features of these quantum states may enable better contrast, resolution, and measurement sensitivity apart from other benefits.

In this chapter, we will discuss the basic Fock/number state, followed by more complex states such as classical coherent state and squeezed quantum states of light. This will be followed by the discussion of the entangled NOON state. Subsequently, we will discuss how these states can be harnessed in optical microscopy for phase super-resolution and precision imaging.

5.1 States of Light

Light as we know, exists in both classical and quantum states. These states (some of them peculiar) offer many possibilities that may be harvested for precision microscopy and imaging. We begin by understanding the basic state of light (Number state) and later on express both classical (Coherent state) and quantum states of light (Squeezed state and entangled state) in terms of basic state.

5.1.1 Fock/Number States

Let us begin with the number/Fock state of light. We start with a single mode field (single frequency, ν) with Hamiltonian H and consider that light is composed of chunks of energy or quanta called photons (of specific frequency). Let the light be in an eigenstate, say, $|n\rangle$, for which the energy eigenvalue is E_n. Here, $|n\rangle$ represents a number state of light in which light field has exactly "n" number of photons/quantas. We use the fundamental postulate of quantum mechanics that relates eigenstate with eigenvalues [1] i.e,

$$H\,|n\rangle = E_n\,|n\rangle \tag{5.1}$$

Using the explicit expression for the Hamiltonian for an optical field i.e., $\hbar\nu\ (\hat{a}^{\dagger}\hat{a} + 1/2)$, the energy eigen-value equation 5.1 becomes,

$$\hbar\nu\left(\hat{a}^{\dagger}\hat{a} + \frac{1}{2}\right)|n\rangle = E_n\,|n\rangle \tag{5.2}$$

where, $\hat{a}$ and $\hat{a}^{\dagger}$ are the annihilation and creation operators. The explicit expressions for the operators can be found in Ref. [2].

Now, applying the operator $\hat{a}$ from left and using the commutator relation, $[\hat{a}, \hat{a}^{\dagger}] = \hat{a}\hat{a}^{\dagger} - \hat{a}^{\dagger}\hat{a} = 1$, gives,

$$\hbar\nu \left(\hat{a}\hat{a}^{\dagger} + \frac{1}{2}\right) \hat{a}\,|n\rangle = E_n\,\hat{a}|n\rangle$$

$$\hbar\nu \left(\hat{a}^{\dagger}\hat{a} + \frac{1}{2}\right) \hat{a}\,|n\rangle = (E_n - \hbar\nu)\,\hat{a}|n\rangle$$

$$H\,\hat{a}|n\rangle = (E_n - \hbar\nu)\,\hat{a}|n\rangle \tag{5.3}$$

Hence, $\hat{a}|n\rangle$ is a state with energy $\hbar\nu$ (or one photon) less than that for the state $|n\rangle$. So representing the new state as,

$$\hat{a}|n\rangle = \alpha|n-1\rangle \tag{5.4}$$

with energy, $(E_n - \hbar\nu) = E_{n-1}$ produces,

$$H\,\hat{a}|n\rangle = E_{n-1}\,\hat{a}|n\rangle \;\; or, \;\; H\,|n-1\rangle = E_{n-1}\,|n-1\rangle \tag{5.5}$$

Now, repeating the process of using the operator $\hat{a}$ and using the commutator relation, ultimately leads to the following relation,

$$H\,\hat{a}|0\rangle = (E_0 - \hbar\nu)\,\hat{a}|0\rangle \tag{5.6}$$

Here, E_0 is the energy of the ground state and since energy cannot be less than the ground state, so the only way the above equation is satisfied is by concluding that,

$$\hat{a}\,|0\rangle = 0 \tag{5.7}$$

where, the state $|0\rangle$ represents a vacuum state.

To obtain the ground state energy, we use the above Eq. 5.7 in the energy eigen-value equation 5.2,

$$E_0 = \frac{1}{2}\,\hbar\nu \tag{5.8}$$

Using the energy value of the ground state E_0, we get the following iterative relation,

$$E_n = E_{n-1} + \hbar\nu = E_0 + n\hbar\nu = \left(n + \frac{1}{2}\right)\hbar\nu \tag{5.9}$$

Now, substituting Eq. 5.9 in Eq. 5.2 leads to,

$$\hat{a}^{\dagger}\hat{a}\;|n\rangle = n\;|n\rangle \tag{5.10}$$

where, $\hat{N} = \hat{a}^{\dagger}\hat{a}$ is known as the number state with an eigen value n, giving the number of quantas in the light field.

Next, we proceed to determine the explicit expression for the number field and express it in terms of vacuum field. To do so, we use the normalization condition and reduce the number state recursively. We begin using the normalization condition for state $|n\rangle$ i.e,

$$\langle n|n\rangle = 1 \tag{5.11}$$

Using the fact that, $\langle n|\hat{a}^{\dagger}\hat{a}|n\rangle = |\alpha|^2\langle n-1|\hat{a}^{\dagger}\hat{a}|n-1\rangle$, and $\langle n|\hat{a}^{\dagger}\hat{a}|n\rangle = n\;\langle n|n\rangle$, we get,

$$\frac{|\alpha|^2}{n}\;\langle n-1|n-1\rangle = 1 \tag{5.12}$$

Noting that the number states are normalized i.e, $\langle n-1|n-1\rangle = 1$, we get,

$$\alpha = \sqrt{n} \tag{5.13}$$

Using Eq. 5.13 in Eq. 5.4 produces,

$$\hat{a}\;|n\rangle = \sqrt{n}\;|n-1\rangle \tag{5.14}$$

Repeating the same process from Eqs. 5.3 to 5.14 for $\hat{a}^{\dagger}$ produces the following relation,

$$\hat{a}^{\dagger}\;|n\rangle = \sqrt{n+1}\;|n+1\rangle \tag{5.15}$$

The above set of equations represents *annihilation* and *creation* operations that result in the annihilation and creation of *photon quanta*.

In addition, the recursive use of the above operator (backwards) produces the following explicit relation connecting the number state $|n\rangle$ and vacuum state $|0\rangle$ i.e.,

$$|n\rangle = \frac{(\hat{a}^{\dagger})^n}{\sqrt{n!}}\;|0\rangle \tag{5.16}$$

This completes the brief description of a single-mode radiation field called the Fock state or photon number state. In short, the field contains an exact number of photons in each mode ν, and the field can be expressed as a statistical mixture of number states. Henceforth, all other states will be expressed in terms of Fock state.

5.1.2 Coherent State

Coherent states are one of the most common states of light. The state contains random number of photons distributed in time, with a constant mean photon flux (say, $|\alpha|^2$) [2]. Accordingly, coherent state of a radiation field is defined as,

$$|\alpha\rangle = e^{-|\alpha|^2/2} \sum_{n=0}^{\infty} \frac{\alpha^n}{\sqrt{n!}} |n\rangle \tag{5.17}$$

Now using Eq. 5.16, the coherent state may be expressed as,

$$|\alpha\rangle = e^{-|\alpha|^2/2}\, e^{\alpha \hat{a}^\dagger}\, |0\rangle \tag{5.18}$$

Noting that, $e^{-\alpha^* \hat{a}}|0\rangle = 0$, we get the following cohesive form,

$$|\alpha\rangle = e^{-|\alpha|^2/2}\, e^{\alpha \hat{a}^\dagger}\, e^{-\alpha^* \hat{a}}\, |0\rangle = D(\alpha)\, |0\rangle \tag{5.19}$$

Using **Baker-Hausdorff formulation,**[1] we can express the displacement operator $(D(\alpha))$ as,

$$D(\alpha) = e^{\alpha \hat{a}^\dagger - \alpha^* \hat{a}}. \tag{5.20}$$

The above relations (Eqs. 5.19 and 5.20) state that a coherent state can be obtained from a vacuum state by applying a displacement operator.

Next, we proceed to determine the probability of finding n number of photons in a single mode coherent field. This is obtained from the modulus-square of the inner product of $|n\rangle$ and $|\alpha\rangle$ i.e,

$$P(n) = |\langle n|\alpha\rangle|^2 = \langle n|\alpha\rangle^* \langle n|\alpha\rangle = \langle \alpha|n\rangle\, \langle n|\alpha\rangle \tag{5.21}$$

Using explicit expression for the coherent state, $|\alpha\rangle$ (Eq. 5.17), the above expression can be simplified to,

$$\begin{cases} \langle n|\alpha\rangle = e^{-|\alpha|^2/2}\, \langle n|\, \sum_{n=0}^{\infty} \frac{\alpha^n}{\sqrt{n!}}\, |n\rangle = e^{-|\alpha|^2/2}\, \langle [\langle n|0\rangle + \alpha\langle n|1\rangle + \ldots + \langle n|\frac{\alpha^n}{\sqrt{n!}}|n\rangle] \\ \langle \alpha|n\rangle = e^{-|\alpha|^2/2}\, \langle n|\, \sum_{n=0}^{\infty} \frac{\alpha^{*n}}{\sqrt{n!}}\, |n\rangle = e^{-|\alpha|^2/2}\, \langle [\langle n|0\rangle + \alpha^*\langle n|1\rangle + \ldots + \langle n|\frac{\alpha^{*n}}{\sqrt{n!}}|n\rangle] \end{cases} \tag{5.22}$$

[1] If $\hat{A}$ and $\hat{B}$ are two operators that satisfy, $[[\hat{A}, \hat{B}],\ \hat{A}] = [[\hat{A}, \hat{B}],\ \hat{B}] = 0$, then, $e^{\hat{A}+\hat{B}} = e^{-[\hat{A},\hat{B}]/2}\, e^{\hat{A}}\, e^{\hat{B}}$.

Noting that, the Fock states are complete orthonormal state,[2] only the nth-term survives, producing,

$$P(n) = \frac{e^{-|\alpha|^2} (\alpha^*\alpha)^n}{n!} = \frac{e^{-\langle n \rangle} \langle n \rangle^n}{n!} \tag{5.23}$$

where, $\langle n \rangle = \alpha^*\alpha = |\alpha|^2$ is the mean.

Thus, a coherent field obeys Poisson distribution with $\langle n \rangle$ as the average number of photons in it. Next, we proceed to explore quantum states of light such as a squeezed state that has implications in high-resolution imaging. We set to explore this in the subsequent sections.

5.2 Squeezed State of Light

In this section, we will discuss one of the unique states of light called squeeze state, and use this light to enable a new kind of microscopy.

As the name suggests, squeezed states are squeezed versions of coherent states (both ground state, ϕ_0 and higher states, ϕ_n with $n > 0$). Such states are key to detect signals below the classical short noise limit. This limits the capability of classical microscopy techniques that use coherent states. Squeezed states are thus a way to get around the short noise limit and detect signals (such as gravitational waves) buried under short noise.

We begin with the Hamiltonian of squeeezed state. A detailed description can be found in [2]. Accordingly, the Hamiltonian can be expressed as,

$$\hat{H} = i\hbar(g\hat{a}^{\dagger 2} - g^*\hat{a}^2) \tag{5.24}$$

where, $\hat{a}^\dagger$ and $\hat{a}$ are the creation and annihilation operators, respectively.

The corresponding squeeze operator is given by,

$$S(\xi) = e^{(\xi^2\hat{a}^2 - \xi^{*2}\hat{a}^{\dagger 2})/2} \tag{5.25}$$

where, $\xi = r\, e^{i\theta}$ is a complex number.

The squeeze operator acting on a coherent state gives rise to squeezed coherent state i.e,

$$|\alpha, \xi\rangle = S(\xi)\, |\alpha\rangle \tag{5.26}$$

[2] Orthonormality condition of Fock states, $\langle n|m \rangle = \begin{cases} 0, & n \neq m \\ 1, & n = m \end{cases}$

Using Eq. 5.19, we can rewrite the above equation in terms of zero-photon/vacuum state,

$$|\alpha, \xi\rangle = S(\xi)\, D(\alpha)\, |0\rangle \tag{5.27}$$

where, $\alpha = |\alpha|\, e^{i\varphi}$ is a complex constant.

To see the effect of squeeze operator, we apply it on a single mode light for which the electric field is given by,

$$E(t) = E_0\, (\hat{a}\, e^{-i\nu t} + \hat{a}^{\dagger}\, e^{i\nu t}) \tag{5.28}$$

where, the operators obey the commutator relation, $[\hat{a}, \hat{a}^{\dagger}] = 1$.

Introducing dimensionless quantities for the position and momentum operators i.e,

$$X' = (\hat{a} + \hat{a}^{\dagger})/2, \qquad P' = (\hat{a} - \hat{a}^{\dagger})/2i \tag{5.29}$$

we get the following commutator relation,

$$[X', P'] = \frac{i}{2} \tag{5.30}$$

Using Eqs. 5.28 and 5.29, the single mode field can be expressed as,

$$E(t) = 2E_0[X' \cos(\nu t) + P' \sin(\nu t)] \tag{5.31}$$

where, X' and P' are Hermition that represents the amplitude of two quadratures of the field with a phase-difference, $\pi/2$.

Next we proceed to find the expectation values of the rotated complex amplitudes (uncertainty) i.e,

$$\begin{cases} \Delta X^{''} = \sqrt{\langle X^{''2}\rangle - \langle X^{''}\rangle^2} \\ \Delta P^{''} = \sqrt{\langle P^{''2}\rangle - \langle P^{''}\rangle^2} \end{cases} \tag{5.32}$$

where, the rotated complex amplitudes are related to the normalized coordinates ($X^{''} = X^{'}\, e^{i\theta}, \quad P^{''} = P^{'}\, e^{i\theta}$). Specifically, considering a rotation of $\theta/2$, we have the following relation,

$$\begin{cases} \Delta X^{''} + i\Delta P^{''} = (X^{'} + iP^{'})\, e^{-i\theta/2} = \hat{a}\, e^{-i\theta/2} \\ \Delta X^{''} - i\Delta P^{''} = (X^{'} - iP^{'})\, e^{-i\theta/2} = \hat{a}^{\dagger}\, e^{i\theta/2} \end{cases} \tag{5.33}$$

From the above Eq. 5.33, we deduce the rotated normalized variables,

$$\begin{cases} \Delta X'' = (\hat{a}\ e^{-i\theta/2} + \hat{a}^{\dagger}\ e^{i\theta/2})/2 \\ \Delta P'' = (\hat{a}\ e^{-i\theta/2} - \hat{a}^{\dagger}\ e^{i\theta/2})/2 \end{cases} \tag{5.34}$$

Using Eq. 5.32, the uncertainty in the position and momentum variable for a squeezed light source is given by,

$$\begin{cases} \Delta X'' = \frac{1}{2}\left[\langle \alpha, \xi|(\hat{a}\ e^{-i\theta/2} + \hat{a}^{\dagger}\ e^{i\theta/2})^2|\alpha, \xi\rangle - \langle \alpha, \xi|(\hat{a}\ e^{-i\theta/2} + \hat{a}^{\dagger}\ e^{i\theta/2})^2|\alpha, \xi\rangle^2\right]^{1/2} \\ \Delta P'' = \frac{1}{2}\left[\langle \alpha, \xi|(\hat{a}\ e^{-i\theta/2} - \hat{a}^{\dagger}\ e^{i\theta/2})^2|\alpha, \xi\rangle - \langle \alpha, \xi|(\hat{a}\ e^{-i\theta/2} - \hat{a}^{\dagger}\ e^{i\theta/2})|\alpha, \xi\rangle^2\right]^{1/2} \end{cases} \tag{5.35}$$

Simplification is possible if we use the following established identities [3],

$$\begin{cases} \langle \hat{a} \rangle = \alpha\ \cosh(r) - \alpha^*\ e^{i\theta}\ \sinh(r), \\ \langle \hat{a}^2 \rangle = \alpha^2\ \cosh^2(r) + \alpha^{*2}\ e^{i2\theta}\sinh^2(r) - 2|\alpha|^2\ e^{i\theta}\ \sinh(r)\ \cosh(r) \\ \qquad\qquad -e^{i\theta}\ \cosh(r)\ \sinh(r), \\ \langle \hat{a}^{\dagger}\hat{a} \rangle = |\alpha|^2\cosh^2(r) + |\alpha|^2\sinh^2(r) + \sinh^2(r) - \alpha^{*2}\ e^{i\theta}\ \sinh(r)\ \cosh(r) \\ \qquad\qquad -\alpha^2\ e^{-i\theta}\ \sinh(r)\ \cosh(r). \end{cases} \tag{5.36}$$

where, r is the squeeze parameter.

Using the above identities, we get the following simplified form,

$$\begin{cases} \Delta X'' = \frac{1}{2}\ e^{-2r} \\ \Delta P'' = \frac{1}{2}\ e^{+2r} \end{cases} \tag{5.37}$$

The above expression for the position and momentum variables 5.37 leads to the following uncertainty relation,

$$\Delta X''\ \Delta P'' = \frac{1}{4} \tag{5.38}$$

This states that the complex amplitude of a squeezed coherent state is an ellipse with similar uncertainty relation to that of coherent state. This is evident from the squeeze parameter, $\eta = |2r|$. However, unlike the coherent state, the squeezed state has the possibility of reduced uncertainty in one conjugate variable at the expense of higher uncertainty in other variable. This translates to the possibility where one variable can be measured with a precision better than the conjugate variable. In the next section, we will use this property of squeezed state to better precision in the measurement of one of the variables.

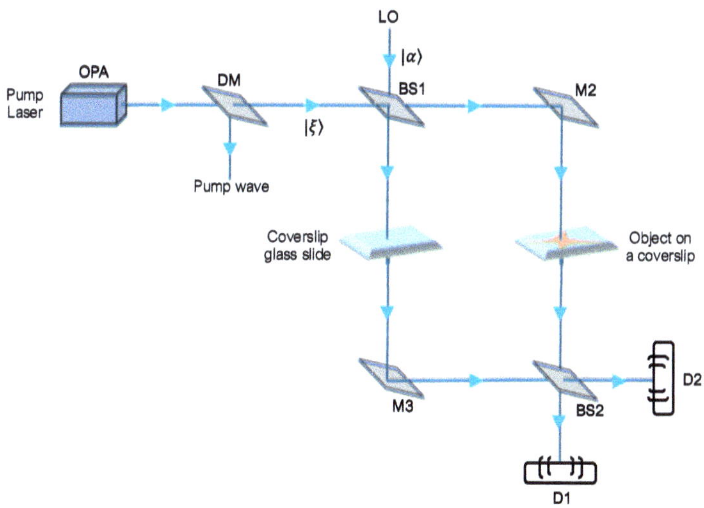

Fig. 5.1 An optical system to detect faint phase objects with sensitivity better that shot-noise limit

5.2.1 Squeezed Light Phase Microscopy

The idea of using squeezed light for optical phase microscopy is first proposed by Kolobov and Kumar in 1993 [4]. This is inspired by the work in interferometry where enhancement of optical measurements beyond short-noise limit is demonstrated [5, 6].

The optical setup of squeezed light microscopy is shown in Fig. 5.1. The system comprises of Mach-Zehnder interferometer with two 50/50 beam-splitters, BS1 and BS2. A plane wave which is also the local oscillator (LO) is injected into one port of BS1, and the other port is injected by squeezed vacuum (originating from OPA). The details for generating squeezed vacuum is detailed in Ref. [4, 7]. A dichroic mirror DM is used to stop the pump wave of OPA from entering the port 2. Light prepared in two different states (coherent and squeezed) are combined by BS1 and propagate two different paths, one path with a glass slide to introduce a constant phase ϕ_0, and sample (say, mouse or human cell on a coverslip) on second path to introduce a slightly different unknown phase. Both the beams then combines at BS2 and gets detected by two different photodetector, D_1 and D_2. The difference in photocurrent density is then observed as function of spatial coordinates,

$$i_0(\rho) = i_1(\rho) - i_2(\rho) \tag{5.39}$$

where, ρ is the function of (x, y) i.e, $\rho(x, y)$.

The next step is the calculation of LO phase (port 1) and the enhancement of phase upon introduction of squeezed light (port 2). Note that, the field entering the first port is the a plane wave which also serves as the local oscillator (LO). The LO field $\hat{a}_1(\rho)$ is in a coherent state with complex amplitude, α and the corresponding quantum state is ψ. This gives the following eigenvalue equation,

$$\hat{a}_1(\rho)\,|\psi\rangle = \alpha\,|\psi\rangle \tag{5.40}$$

The field entering the second port is a multimode squeezed vacuum, $\hat{a}_2(\rho)$. The field is generated by a travelling-wave OPA with vacuum-state input pumped by a monochromatic plane wave discussed by Kobolov and Sokolov [8–10]. Considering, $\hat{a}_1(\rho)$ and $\hat{a}_2(\rho)$ as the photon annihilation operators for the light waves at the two input ports of the the Mach-Zehnder interferometer, with $\hat{a}_i^\dagger(\rho)\,\hat{a}_i(\rho)$ as the photon flux density, the mean photocurrent is given by,

$$\langle i_0(\rho)\rangle = \eta\,[\,\langle \hat{a}_1^\dagger(\rho)\,\hat{a}_1(\rho) - \hat{a}_2^\dagger(\rho)\,\hat{a}_2(\rho)\rangle \cos\phi - \langle \hat{a}_1^\dagger(\rho)\,\hat{a}_2(\rho) + \hat{a}_2^\dagger(\rho)\,\hat{a}_1(\rho)\rangle \sin\phi\,] \tag{5.41}$$

where, η is the quantum efficiency of the photodetector, and ϕ is the phase-difference for waves in both the arms.

Following Kolobov, Sokolov and Kumar [4, 8], the noise spectrum of the photocurrent density between LO field and squeeze field is,

$$\Delta i_0^2(q,\theta) = \langle i\rangle\,[\,1-\eta+\eta(e^{2r(q)}\,\cos^2\theta(q) + e^{-2r(q)}\,\sin^2\theta(q)\,)] \tag{5.42}$$

where, q is the spatial frequency, θ is the relative phase between LO and squeeze light.

Introducing the phase difference ϕ by invoking Eq. 5.41 in the noise spectrum term (Eq. 5.42) of the interferometer produces [11],

$$\begin{aligned}\Delta i_0^2(q,\theta,\phi) = \eta\,|\alpha|^2[\,1-\eta+\eta(e^{2r(q)}\,\cos^2\theta(q) + e^{-2r(q)}\,\sin^2\theta(q)\,)\,\sin^2\phi \\ +\eta\,\cos^2\phi]\end{aligned} \tag{5.43}$$

where, $\langle i\rangle = \eta\,|\alpha|^2$.

Now, considering that the LO field is strong as compared to squeezed-vacuum field, only the first term survives in Eq. 5.41 giving,

$$\langle i_0(\rho)\rangle = \eta\,[\,\langle \hat{a}_1^\dagger(\rho)\,\hat{a}_1(\rho)\cos\phi\,] = \eta|\alpha|^2\cos\phi \tag{5.44}$$

Note that the phase sensitivity and minimum detectable phase change are defined as,

$$\begin{cases} S(\phi) = \left|\frac{\partial\langle i_0\rangle}{\partial\phi}\right| \\ \Delta\phi_{min}^2 = \frac{1}{S_{max}^2}\,\Delta i_0^2 \end{cases} \tag{5.45}$$

The interferometer has maximum sensitivity for phase-difference $\phi = (2n+1)\frac{\pi}{2};\ n = 0, 1, 2, \ldots$. At these points, the sensitivity and minimum detectable phase change are,

$$\begin{cases} S(\phi)|_{max} = \eta\ |\alpha|^2 \\ \Delta\phi^2_{min} = \frac{1}{\eta|\alpha|^2}\left[\ 1 - \eta + \eta\{\cos^2\theta(q)\ e^{2r(q)} + \sin^2\theta(q)\ e^{-2r(q)}\}\right] \end{cases} \tag{5.46}$$

Now for $\theta(q) = \pm\pi/2$ and quantum efficiency, $\eta \sim 1$, the minimum detectable phase is,

$$\Delta\phi^2_{min} = \frac{1}{|\alpha|^2}\ e^{-2r} \tag{5.47}$$

So, the introduction of squeeze vacuum state in port 2 of the interferometer has scaled the minimum detectable phase by e^{-2r}, whereas if port 2 is blocked, the phase would be, $\Delta\phi^2_{min} = \frac{1}{|\alpha|^2}$. This means that, squeezed vacuum states reduce the minimum detectable phase beyond the classical limit.

To realize the implication of minimal detectable phase, a simple imaging experiment (similar to Fig. 5.1) can be carried out with sample in one arm, and a flat glass block in the other arm (see, Fig. 5.1). In principle, the sample can be scanned by a piezo-based scanning stage to obtain phase information point-by-point, which can then be put together to obtain phase map of the sample object better than classical limit. Overall, this enhances our ability towards precision phase microscopy.

5.3 Phase Super-Resolution Microscopy using Entangled NOON States

As the name suggests, super-resolution is enabled by entangled NOON quantum state. We will begin with NOON states before going for N-particle entangled photon microscopy [12].

A NOON state is an entangled state which is a superposition of all the particles simultaneously occupying a quantum state. The state may be expressed as [13],

$$|\psi_{NOON}\rangle = \frac{1}{\sqrt{2}}\left[|N, 0\rangle_{a,b} + |0, N\rangle_{a,b}\right] \tag{5.48}$$

where, $|N, 0\rangle_{a,b}$ is a state where all the N particles in mode a occupy one of the two states and 0 particles in mode b of the remaining state [14]. Here, Eq. 5.48 describe a superposition of two distinct Fock/Number states, $|N, 0\rangle_{a,b} = |N\rangle_a\ |0\rangle_b$ and $|0, N\rangle_{a,b} = |0\rangle_a\ |N\rangle_b$.

Under the same Hamiltonian ($\hat{H} = b^{\dagger}\hat{b}$), the NOON state evolves to,

$$|\psi_{NOON}\rangle = \frac{1}{\sqrt{2}}\left[|N, 0\rangle_{a,b} + e^{iN\varphi}|0, N\rangle_{a,b}\right] \tag{5.49}$$

Now, consider the relative phase observable, $\hat{\phi}_N = |0, N\rangle\langle N, 0| + |N, 0\rangle\langle 0, N|$. We wish to measure the observable i.e,

$$\begin{aligned}\langle\, \hat{\phi}_N \,\rangle &= \langle\psi_{NOON}|\, \hat{\phi}_N \,|\psi_{NOON}\rangle \\ &= \frac{1}{2}[e^{iN\varphi} + e^{-iN\varphi}] \\ &= \cos(N\varphi)\end{aligned} \tag{5.50}$$

The variance/uncertainty in the measured phase observable is given by,

$$\begin{aligned}\Delta\hat{\phi}_N &= \sqrt{\langle\hat{\phi}_N^2\rangle - \langle\hat{\phi}_N\rangle^2} \\ &= \sqrt{1 - \cos^2(N\varphi)} \\ &= \sin(N\varphi)\end{aligned} \tag{5.51}$$

So, the minimum detectable phase is,

$$\Delta\varphi_{NOON} = \frac{\Delta\hat{\phi}_N}{\left|\frac{\partial\langle\hat{\phi}_N\rangle}{\partial\varphi}\right|} = \frac{1}{N} \tag{5.52}$$

Classical Case Now, consider N copies of a single particle superposition state i.e,

$$|\psi\rangle = \frac{1}{\sqrt{2}}\left[|1, 0\rangle_{a,b} + |0, 1\rangle_{a,b}\right] \tag{5.53}$$

where, $|1, 0\rangle_{a,b}$ is a state where all the single particles in mode a occupying one of the two states and 0 particles in mode b of the other state.

The state will evolve to the state,

$$|\psi\rangle = \frac{1}{\sqrt{2}}\left[|1, 0\rangle_{a,b} + e^{i\phi}\,|0, 1\rangle_{a,b}\right] \tag{5.54}$$

Now, the measurement on the phase observable, $\hat{\phi} = |0, 1\rangle\langle 1, 0| + |1, 0\rangle\langle 0, 1|$ for a total of N times produces,

$$\langle\hat{\phi}\rangle = N\langle\psi|\hat{\phi}|\psi\rangle = N\, \cos\varphi \tag{5.55}$$

The variance is given by,

$$\Delta\hat{\phi} = \sqrt{\langle\hat{\phi}^2\rangle - \langle\hat{\phi}\rangle^2} = \sqrt{N - N\cos^2\varphi} = \sqrt{N}\sin\varphi \tag{5.56}$$

So, the minimum detectable phase is,

$$\Delta\phi_{class} = \frac{1}{\sqrt{N}} \tag{5.57}$$

So, comparing Eqs. 5.52 and 5.57 gives, $\frac{\Delta\phi_{NOON}}{\Delta\phi_{class}} = \frac{1}{\sqrt{N}}$.

A comparison of single particle state (classical case) and N-particle entangle state (quantum case) reveals $\sqrt{N}$-times better phase measurement for quantum N-particle state as compared to the classical single particle state. This is a purely quantum mechanical effect as a result of quantum entanglement. An optical microscopy system employing such states are capable of phase super-resolution imaging.

References

1. Mondal PP (2022) Book: classical & quantum optics—chapter 5. American Institute of Physics publication, Melville
2. Mondal PP (2022) Book: classical and quantum optics. American Institute of Physics Publishers, New York
3. Scully MO, Zubairy MS (1997) Book: quantum optics. Cambridge University, Cambridge
4. Kolobov MI, Kumar P (1993) Sub-shot-noise microscopy: imaging of faint phase objects with squeezed light. Opt Exp 18:849–851
5. Xiao M, Wu L-A, Kimble HJ (1987) Phys Rev Lett 59:278
6. Grangier P, Slusher RE, Yurke B, LaPorta A (1987) Phys Rev Lett 59:2153
7. Aytir O, Kumar P (1992) Opt Lett 17:529; Kim C, Li R-D, Kumar P (1993) In: Quantum Electronics Laser Science Conference. 1993 OSA Technical Digest Series, vol. 12. Optical Society of America, Washington, paper QThF1
8. Kolobov MI, Sokolov IV (1989) Sov Phys JETP 69:1097
9. Kolobov MI (1991) Phys Rev A 44:1986
10. Sokolov IV (1991) Sov Phys JETP 73:421
11. Kok P (2002) Creation of large-photon-number path entanglement conditioned on photodetection. Phys Rev A 65:052104
12. Mitchell MW et al (2004) Super-resolving phase measurements with a multiphoton entangled state. Nature 429:161–164
13. Sanders BC (1989) Quantum dynamics of the nonlinear rotator and the effects of continual spin measurement. Phys Rev A 40:2417
14. Bennett L et al (2023) NOON state measurement probabilities and outcome fidelities: a Bethe ansatz approach. J Phys A Math Theor 56:505202

Introduction to Molecular Physics and Fluorescence Spectroscopy

6

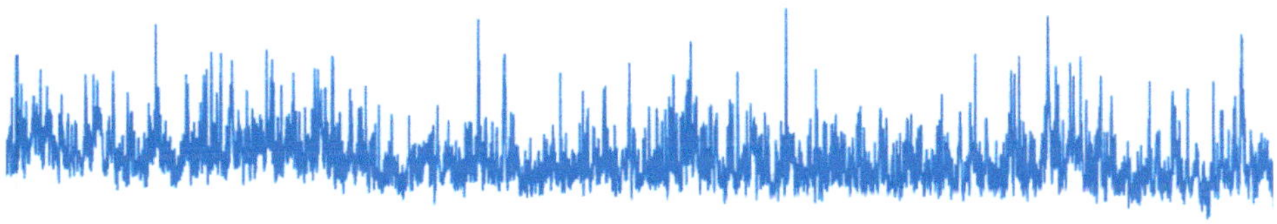

The chemistry behind formation of molecules is the basis for most of the optical spectroscopy, fluorescence microscopy and optical imaging. This involves electronic and vibronic states of molecules that gives rise to specific spectra as its fingerprint. So, the structure of the molecule along with the transitions involving electronic states plays crucial role. This leads to an important class of theory for approximate understanding of molecules called molecular orbital theory (MOT). In this chapter, we will discuss the formation of diatomic and polyatomic molecules along with the selection rules involving electronic states that gives rise to specific spectra including the phenomena of fluorescence. All this will have its base in MOT. Specifically, we will focus on fluorescence and laws governing them including Kasha's law, Stokes shift and Frank-Condon principle. Then we will focus on a few well-known fluorescent dyes/proteins that are frequently used in fluorescence microscopy, and discuss their spectral properties based on MOT. Specifically, here we concentrate on photoactivable fluorophores that are key to super-resolution microscopy and spectroscopy.

Molecular spectroscopy is at the heart of modern imaging techniques that involve infrared, Raman and fluorescence. In this chapter, we will discuss the basics of molecule, its structure and electronic states that ultimately leads to fluorescence based spectroscopy/imaging including single molecule imaging. So, we will begin with molecular spectroscopy and go on to understand molecules based on pictorial representation of Molecular Orbital Theory (MOT).

P. P. Mondal, S. Hess, *Single-Molecule Super-Resolution Optical Microscopy*,
https://doi.org/10.1007/978-3-032-13442-4_6

6.1 Atomic and Molecular Spectroscopy

Most of the spectroscopy involving molecules require electromagnetic radiation as a probe. This is due to the broad range of electromagnetic spectrum where molecules are active. Specifically, the spectrum (from lower to higher wavelength) can be divided into 5 prominent regimes as discussed below:

X-ray Region This regime spans from few hundred picometers to few nanometers (wavelength in the range, [100 pm–10 nm]). The light-matter/light-molecule interaction in this spectral regime predominantly involve inner-electrons of an atom/molecule. The corresponding energy involved are in the range of 10^4 kiloJoules/mole. Some of the commonly known investigation techniques that use X-ray as a probe are, X-ray diffraction (XRD) in basic research and computed tomography (CT) in medical diagnosis.

Visible and UV Region This is the most common and directly observable regime for which the wavelength ranges from 100 nm–1 μm. The interaction in this regime is mostly electronic in nature and mostly outermost valence electrons are involved with energy difference in the range of few hundred kilojoules/mole. Prominent techniques that use this regime are, microscopy, optical spectroscopy, cytometry and optical astronomy.

Infrared Region This regime is one of the extensively used EM regime for understanding molecular bonds and overall chemistry of molecules. The regime spans from $\sim$ [1–100 μm] and the interactions involved are vibrational levels of molecules. The techniques using this regime are vibrational spectroscopy and Raman spectroscopy, that are used in fundamental research, food, industry and pharmaceutical sciences.

Microwave Region The span of this regime has a large spectrum beginning from few hundred micrometer to about a centimetre i.e, [100 μm–1 cm]. The regime involves the rotational levels of molecules, and the energy difference involved are of the order of few hundred joules/mole. The techniques that use this regime are rotational spectroscopy, weather forecasting and communication.

Radiofrequency Region The wavelengths involved in this regime are in the range, [1 cm–10 m]. The interaction in this case is predominantly due to reversal of nuclear/electron spin of a molecule and the corresponding energy-change is of the order of 0.01–10 Joules/mole. The respective investigation techniques are called nuclear/electron spin resonance (nmr/esr) used in medical science and radio-astronomy.

Table 6.1 sums-up the regions of electromagnetic spectrum with their mechanism of action along with known applications. Here, we are mostly interested in visible and near-IR region of which fluorescence ([400–980 nm]) is a small spectrum. This is predominantly due to the advances made in recent decade and the promise it holds in understanding biological/physical processes with the ultimate single molecule precision.

Table 6.1 Classification of electromagnetic spectrum based on mechanism of action

	X-ray Distribution	Visible & UV	Infrared (IR)	Microwave	Radio-frequency
Wavelength	10 nm–100 pm	1 μm–100 nm	1–100 μm	100 μm–1 cm	1 cm–10 m
Frequency	[3×10^{16}–3×10^{18} Hz]	[3×10^{14}–3×10^{16} Hz]	[3×10^{12}–3×10^{14} Hz]	[3×10^{10}–3×10^{12} Hz]	[3×10^{6}–3×10^{10} Hz]
Difference Energy	$\sim 10^4$ kJ/mole	$\sim$ 100 kJ/mole	$\sim$ 10 kJ/mole	$\sim$ 100 J/mole	10^{-3}–10 J/mole
Action Mechanism	Change of inner electron distribution	Change of valence electron distribution	Change of molecular vibrations	Change of molecular rotations	Change of spin
Applications	XRD, CR etc.	Microscopy, Spectroscopy etc.	Raman Spec., Drug Discov. etc.	Spectroscopy, Comm. etc.	NMR, ESR, fMRI etc.

6.2 The Molecular Orbital Theory (MOT)

This is probably the simplest theory of molecules that can predict single molecule properties (both as single entity and in ensemble) without sacrificing scientific rigor. Several theories have been suggested to understand the formation of molecules from atoms, but the complexity involved makes it almost impossible even for simple diatomic molecules. MOT is probably the simplest of all and provides pictorial representation of molecule formation. The theory is particularly suited for understanding electronic transitions and have a good degree of similarity with the well-known atomic structure. This appears to be the most apppropriate theory to understand fluorescent molecules that largely involve electronic transitions. Particularly, MOT aids to a better understanding of the selection rules and transitions between molecular orbitals in photoswitchable/photoactivable molecules that are frequently used in single molecule super-resolution microscopy. In this section, we embark on a journey to understand the structure of molecules and how it determines its spectral properties.

The basics of quantum mechanics teaches us that energy states in a closed system such as single molecule/atom absorb and emit specific wavelengths of light. This directly implies that unlike classical world of continuous energy, molecules have discrete energy levels, Upon irradiation with a visible light, selective transitions takes place between molecular orbitals (that describes distribution of electrons in molecules) with well-defined energy levels. Mathematically, these orbitals are expressed by a three-dimensional wave function (or orbital, ψ) and the corresponding energy level can be calculated from the celebrated Schrodinger equation.[1] In specific, these orbitals depends on three independent quantum numbers, n, l, m, and rules (similar to that of atomic orbitals,[2,3,4]) apply on how these orbitals are occupied in the electronic states and which transitions can take place. These principles ultimately lead to the *selection rules*.

Molecule formation reduces the overall energy of the system and thereby brings in more stability. MOT supposes atoms approach to form a molecule, where individual entity of atomic orbitals disappears leading to the formation of molecular orbitals for the conjugate system (the molecule); the shape and energy of these orbitals ($\psi_{n,l,m}$) are calculated using Schrodinger equation (in terms of 3 quantum numbers). The shape of the molecular orbital is defined as the space where the electrons belonging to that orbital spends most of its time, meaning measurement (if practically possible) will have a very high probability of finding electrons in that space. Since detailed calculations using Schrodinger equation is

[1] Time-independent Schrodinger equation: $-\frac{\hbar^2}{2m}\nabla^2\psi = E\psi$.

[2] **Aufbau Principle:** Filling of electrons must take place starting from lowest energy levels.

[3] **Pauli Exclusion Principle:** No two electrons in the same orbital can have the same set of quantum numbers (n, l, m).

[4] **Hund's Rule:** The electrons singly occupy orbitals until the sublevel is filled before they double-up.

extremely complicated and may not have a closed form, MOT gives an alternate pictorial representation. Here we begin with the basic postulates of MOT:

6.2.1 Postulates of MOT

1. Molecular orbitals (say, $\psi_{mo\pm}$) can be constructed from the atomic orbitals (say, ψ_1 and ψ_2) of constituent atoms forming the molecule.
2. Molecular orbitals are obtained by linear combination (both sums and differences) of atomic orbitals (LCAO). In general, for n atomic orbitals combining to form a molecule, the ith molecular orbital is given by,

$$\psi_{mo,i} = \sum_n c_{ni}\psi_n \tag{6.1}$$

   where, the coefficients c_{ni} are the contributions of n atomic orbitals, and can be determined using Hartee-Fock method. Here, $|\psi_{mo\pm}|^2$ is the probability of finding the electron in space.
   For example, the molecular orbitals of a diatomic molecule would be,

$$\begin{cases} \psi_{mo+} = c_1\psi_1 + c_2\psi_2 \\ \psi_{mo-} = c_1\psi_1 - c_2\psi_2 \end{cases} \tag{6.2}$$

3. The orbitals thus formed are best classified as bonding and antibonding depending upon the constructive and destructive overlap of constituent atomic orbitals.

With these postulates as the starting point, we move forward to understand diatomic molecules and the newly formed molecular orbitals arising out of constituent atoms.

The first thing we note is that the molecular orbitals resemble atomic orbitals close to one of the constituent atoms. Considering diatomic molecules (say, Hydrogen H_2), each consisting of $1s$ orbitals (ψ_{1s}) can give rise to two molecular orbitals upon combination,

$$\begin{cases} \psi_{1s\sigma} = \psi_{1s}^{(1)} + \psi_{1s}^{(2)} \\ \psi_{1s\sigma*} = \psi_{1s}^{(1)} - \psi_{1s}^{(2)} \end{cases} \tag{6.3}$$

where, $\psi_{1s}^{(1)}$ and $\psi_{1s}^{(2)}$ are respectively the atomic orbitals of individual H-atoms (marked as, 1 and 2). It may be noted that, it is hard to distinguish which is which, i.e, one can't distinguish between H-atoms.

A pictorial representation of bonding and anti-bonding orbitals formed from the combination of two atomic orbitals is shown in Fig. 6.1. It is imperative to note that, the constructive combination of two orbitals lead to an increase in the region of orbital overlap, giving it an ellipsoid shape. Similarly, a destructive combination of two orbitals

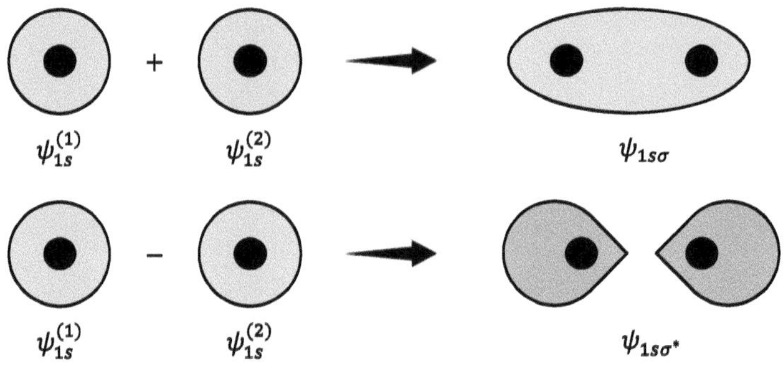

Fig. 6.1 Formation of molecular orbitals ($\psi_{1s\sigma*}$ and $\psi_{1s\sigma}$) from two atomic orbitals ($\psi_{1s}^{(1)}$ and $\psi_{1s}^{(2)}$)

lead to decrease in the value of molecular orbital in the region of overlap. This can be further confirmed by detailed calculation. As a consequence, $\psi_{1s\sigma*}$ and $\psi_{1s\sigma}$ are called anti-bonding and bonding orbitals, respectively. Another aspect associated with orbitals is inversion which determines symmetry about their internuclear axis. This shows that bonding orbital remains unchanged at the inter-nuclear axis, while anti-bonding orbital changes upon inversion. Hence, bonding and anti-bonding orbitals are also described as even (represented by, g) i.e, $\psi_{1s\sigma_g}$ and odd (represented by u) i.e, $\psi_{1s\sigma_u^*}$, since bonding orbital does not reverse sign whereas anti-bonding orbitals reverses sign upon inversion. This even/odd symmetry considerations arise only for homonuclear molecules (e.g. H_2, O_2 etc.).

For simplicity, we will proceed our discussion with a simple diatomic molecule before understanding polyatomic molecules using MOT. Note that, atomic orbitals can have other type of electronic distribution (such as, $p,\ d,\ f$ etc.) around the nucleus other than the well-known s-type depending upon the symmetry, shape and energy of atomic orbitals. Figure 6.2 shows the molecular orbital energy level diagram for a typical diatomic molecule (other than Hydrogen molecule). Next, we will try to understand few diatomic molecules. For example, consider N_2 molecule, where the individual nitrogen atom have an electronic configuration given by, $1s^2\ 2s^2\ 2p^3$. In total, N_2 molecule formation involves 14 electrons (7 from each Nitrogen atom). Filling the newly formed molecular orbitals require full compliance of fundamental laws (Aufbau principle, Pauli's exclusion principle and Hund's rule). Accordingly, the configuration for N_2 molecule will be, $1s\sigma_g^2\ 1s\sigma_u^{*2}\ 2s\sigma_g^2\ 2s\sigma_u^{*2}\ 2p\sigma_g^2\ 2p\pi_u^4$.[5] Similarly one can determine distribution of electrons for other diatomic molecules as well.

[5] Note that, $s,\ p,\ d$ orbitals can take upto a maximum of $2,\ 6,\ 10$ electrons respectively.

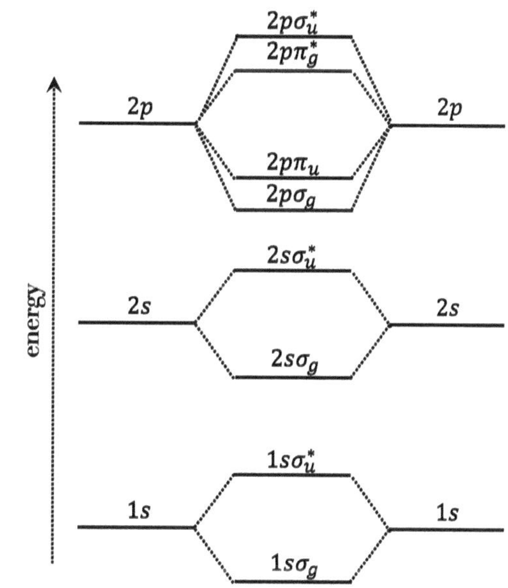

Fig. 6.2 Energy-level diagram showing the formation of molecular orbitals in a typical diatomic molecule from the atomic orbitals of constituent atoms

6.2.2 Diatomic Molecules, Their Spectrum & Selection Rules

In this section, we will discuss the energy states of a simple diatomic molecule, the selection rules that comes to play and the resultant spectrum. We take on the example of simple H_2 molecule and determine the ground and few excited states, followed by the transitions that may arise between the states.

The H_2 molecule contain two electrons, one from each H-atom. In the ground state, both the electrons will occupy the same lowest orbital i.e, $1s\sigma_g$, and hence must be paired forming a singlet state (as dictated by Pauli's exclusion principle). Since both the electrons are σ electrons (with the axial component of orbital angular momentum, $\lambda = 0$), gives $\Lambda = \sum_i \lambda_i = 0$, and its state as, ${}^1\Sigma$.[6] Further sub-division is normally incorporated based on the plane of symmetry of the wavefunction about the inter-nuclei axis, which are represented by $+$ or $-$ based on unchanged or change of symmetry. So, the ground state of the molecular hydrogen can be expressed in complete form as, $(1s\sigma_g)^2\ {}^1\sum_g^+$. For hydrogen molecule, a lot of excited singlet states exists, noting that a electron in the ground state can be raised to any of the available excited states. Considering promotion to only the non-starred states, possible excited states are, $(1s\sigma_g 2s\sigma_g)$, $(1s\sigma_g 2p\sigma_g)$ and $(1s\sigma_g 2p\pi_u)$.

Excited States

Note that, the state $(1s\sigma_g 2s\sigma_g)$ has both the electrons as σ electrons for which orbital angular momentum $\Lambda = \lambda_1 + \lambda_1 = 0$. Taking only the singlet states, the total spin angular

[6] Note that, $\Lambda = 0$ is represented by Σ.

momentum $S = 0$. In addition, the constituent orbitals are even and symmetrical, giving the complete state as, $(1s\sigma_g 2s\sigma_g)\ {}^1\Sigma_g^+$.

The electrons in the next excited state $(1s\sigma_g 2p\sigma_g)$ has σ electrons, but overall state is odd (u). So, one of the electrons is in the even $1s\sigma_g$ state, whereas the other in odd $2p\sigma_g$ state, giving the combined state as an odd state. So, the combined state can be represented as, $(1s\sigma_g 2p\sigma_g)\ {}^1\Sigma_u^+$.

Next consider the state, $(1s\sigma_g 2p\pi_u)$ that has 1 σ and 1 π electron. The state is not symmetric, making its wavefunction odd and has a total orbital angular momentum of, $\Lambda = \lambda_1 + \lambda_2 = 1$, giving a combined state, $1s\sigma_g 2p\pi_u\ {}^1\Pi_u$. (Here, information related to $\Lambda = 1$ (singlet state) is represented in ${}^1\Pi_u$)

However, the states can be triplet depending on the fact that electron spins are parallel giving a total spin of, $S = 1$. Following Pauli's exclusion principle, both the electrons cannot occupy the same state, so the excited state with the lowest energy will be, $(1s\sigma_g 2s\sigma_g)\ {}^3\Sigma_g^+$, $(1s\sigma_g 2p\sigma_g)\ {}^3\Sigma_u^+$ and $(1s\sigma_g 2p\pi_u)\ {}^3\Pi_u$. Here, $\Lambda = 3$ (triplet state) is represented in ${}^1\Pi_u$. Figure 6.3 shows a pictorial representation of the distribution of electrons for the singlet and triplet states.

Note that, other higher excited states involve $3s$, $3p$, $4s$, $4p$ states with the possibility of d-type orbitals.

Selection Rules

Experimentally, it was found that not all the transitions between the ground and excited states are possible. However, a majority of transitions are dictated by selection rules as discussed below:

1. During transitions, the change in total orbital angular momentum must be follow:

$$\Delta\Lambda = 0,\ \pm 1 \tag{6.4}$$

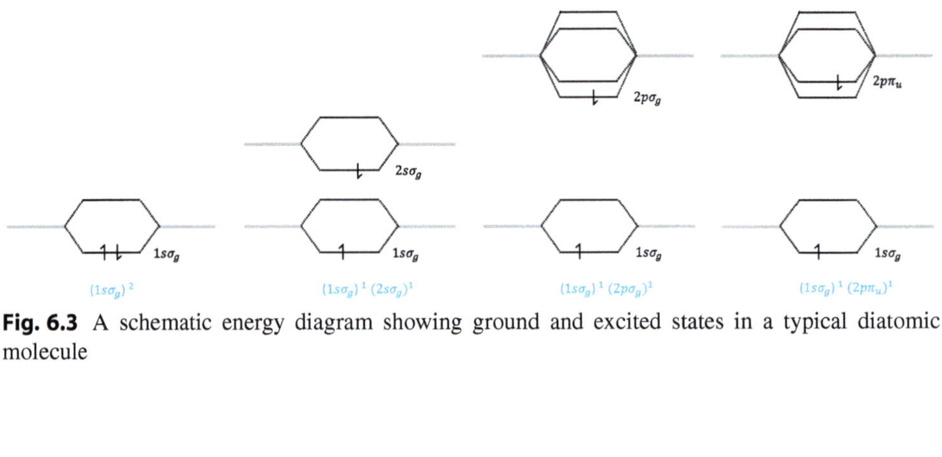

Fig. 6.3 A schematic energy diagram showing ground and excited states in a typical diatomic molecule

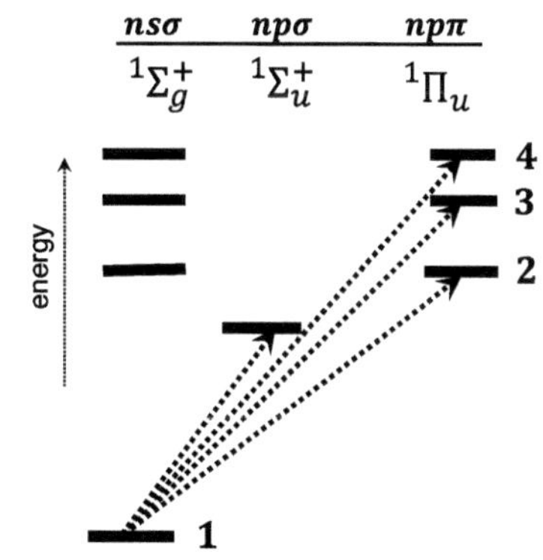

Fig. 6.4 Energy-level diagram showing the formation of molecular orbitals in a typical diatomic molecule from the atomic orbitals of constituent atoms

Accordingly, $\Sigma \leftrightarrow \Sigma$, $\Sigma \leftrightarrow \Pi$ and $\Pi \leftrightarrow \Pi$ are allowed, among other. However, $\Sigma \leftrightarrow \Delta$ is not allowed.[7]

2. The change in total spin quantum number must be zero i.e,

$$\Delta S = 0 \tag{6.5}$$

Accordingly, the transitions between $singlet \leftrightarrow singlet$ states are allowed, however, $singlet \leftrightarrow triplet$ are not allowed.[8]

3. Symmetry changes play a mjor role and accordingly, the following rules apply:

$$\begin{cases} g \leftrightarrow u \\ g \not\leftrightarrow g; \quad u \not\leftrightarrow u \end{cases} \tag{6.6}$$

Accordingly, the transition from Σ^+ into other Σ^+ or Π and Σ^- into other Σ^- or Π are possible.

As an example, some of the allowed transitions from ground state to the lowest excited states is shown in Fig. 6.4. Similar transitions are also allowed to other higher singlet states, but not to the triplet states ($^3\Sigma_g^+$, $^3\Sigma_u^+$, $^3\Pi_u$ etc.) due to the violation of selection rules (see, 6.5).

Now, we are good to go ahead to understand the phenomenon of fluorescence based on molecular orbital theory.

[7] Note that, $\Lambda = 0, 1, 2, \ldots$ are represented by, $\Sigma, \Pi, \Delta, \ldots$, respectively.

[8] Note that, singlet state has a total of 0 and triplet state has a total spin of 1.

6.3 Laws of Fluorescence

The general term is photoluminescence which is the emission of light from a substance exposed to radiation. They can be primarily divided into two major categories: Fluorescence and Phosphorescence. In this section, we will focus on fluorescence and use MOT for better understanding.

In general, fluorescence is the rapid emission (in the order of few nanoseconds) of photons from an excited molecule. In excited state, the electron in the excited orbital is paired opposite to the second electron in the ground-state orbital. As per the selection rules, *singlet* $\rightarrow$ *singlet* transition is allowed by selection rules, so the transition of excited state elections occurs rapidly, The difference energy is emitted in the form of light and the process generally happens in few nanoseconds. The corresponding rate of emission is $\sim 10^8$–$10^9\ s^{-1}$.

On the other hand, phosphorescence is the emission of light from a triplet state. However, it is important to note that emission from excited triplet state to ground singlet state is forbidden by the selection rules, and hence electron takes about milli-to-few seconds time to return to singlet ground state. So, the emission rates are slow, approximately, 10^6 times slower than fluorescence. Note that, quantum mechanics gives a non-zero probability of such events (be it emission from excited singlet to ground singlet state or excited triplet to ground singlet state) to occur. Consequently, forbidden transitions do occur but rate of the process is very slow.

This is better understood based on the simplified pictorial version assisted by molecular orbital theory called Jablonski diagram (see, Fig. 6.5). Accordingly, fluorescence involves transition between singlet ground state (S_0) to excited states (S_1, $S_2, \ldots, S_n$) upon absorption of an appropriate energy photon. The molecule quickly relaxes to the lowest vibrational level of first excited singlet state (S_1). This process is fast (order of few pico-seconds) and termed as called *internal conversion*. Compared to the time scale involved in fluorescence emission (order of few nano-seconds), internal conversion time can be neglected. Hence the system quickly relaxes to the lowest vibrational state of the first excited electronic state before returning to the ground state (to the higher vibrational state

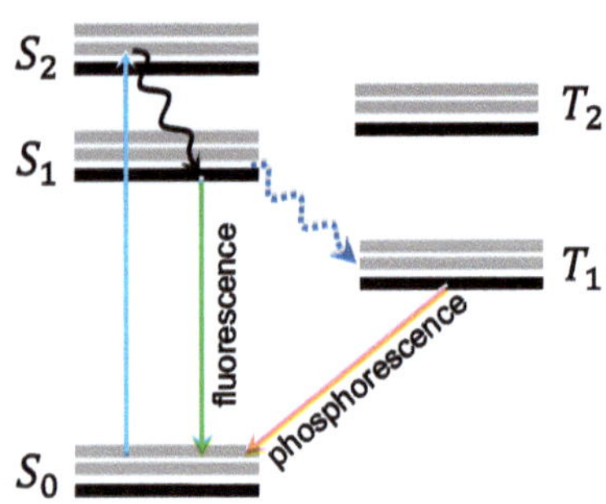

Fig. 6.5 Jablonski diagram showing fluorescence and phosphorescence

of electronic ground state) which then quickly reaches thermal equilibrium. So, the energy is dissipated in the form of fluorescence emission.

This is very different for phosphorescence, where molecule in the excited state (S_1) undergoes spin-conversion and quickly relaxes to the nearby lower energy triplet state (T_1). This process (transition from S_1 to T_1) is called intersystem crossing. Since relaxation to the singlet ground state (S_0) is forbidden by the selection rules, the relaxation process is very slow and emission from triplet states are several orders of magnitude less than fluorescence. However, it is important to note that, present of heavy atoms in a molecule facilitate intersystem crossing (largely due to spin-orbit coupling). All these factors affect emission from single molecules.

Since our interest is central to fluorescence, we will focus our attention on the governing laws/principles of fluorescence as states below:

6.3.1 Stokes' Shift

The fact that a shift of wavelength occurs during fluorescence is discovered by Stokes [1]. In other words, fluorescence typically occurs at higher wavelengths or lower energies. One of the first substance to show fluorescence is quinine solution which strongly absorbs ultraviolet (that we cannot see) and emits blue color light. Subsequent years have seen a similar effect in many other substances including fluorescein.

This shift is primarily due to decay of excited state molecules to the lowest vibrational state of the first excited singlet state, S_1. This is after excitation with a light of higher energy (or small wavelength). After relaxation to the electronic ground state (generally to higher vibrational state), fluorophores further decay to the thermalized vibrational state of the ground state (S_0). In addition, there can be a host of other process that may cause Stokes shift such as, energy transfer and complex formation. Overall, there is a loss of energy in the entire process and the resultant energy (emitted in the form of radiation) is always less than the incident light, causing a shift in the wavelength, $\Delta\lambda = \lambda_{exc} - \lambda_{emi}$.

6.3.2 Kasha's Rule

The next important characteristics of fluorescence is Kasha's rule, which states that, for emission from a single state, the fluorescence emission spectrum is independent of its excitation mechanism. This means that the emission spectra of a fluorophore remains same designating it as a key characteristics of a fluorophore. This is knows as Kasha's rule [2].

Molecules excited to higher electronic states ($S_{n>2}$) can rapidly relax to the lowest vibrational state of the first excited state (S_1). The time scale of this process is in picoseconds as compared to fluorescence that occurs in nano-seconds. This is due to the fact that, there is a strong overlap of numerous excited state molecular orbitals having nearly equal energy, causing rapid relaxation. So, overall this states that the emission spectra is usually independent of excitation wavelength. However, exceptions do exist.

6.3.3 Frank-Condon Principle

This principle states that, a molecule undergoing an electronic transition causes no significant change in nuclear configuration of a molecule. In other words, electronic transitions occur at a time scale much shorter than nuclear motions. Consequently, the vibrational states to and from which absorption and emission occur are those that cause a minimal change in the nuclear coordinates. This process is also termed as vertical transition and corresponding transition probability can be calculated at a fixed nuclear position. This is also the reason why both absorption and emission produce molecules in vibrationally excited states.

As a consequence, this leads to an interesting rule in fluorescence called mirror symmetry of absorption and emission spectra, which can be explained based on both Franck–Condon principle and Kasha's rule in both absorption and fluorescence. In simpler terms, if a particular transition probability (Frank-Condon factor) between two vibrational levels (say, $0-0,\ 0-1,\ 0-2$ etc.) of involved electronic states (S_0 and S_1) is largest in absorption, then the probability of reciprocal transition in emission is equally probable.[9]

6.4 Characteristics of Fluorescence

In this section, we discuss some of the key characteristics of fluorescence phenomena and the interlink between them.

6.4.1 Quantum Yield

The number of photons emitted relative to that of absorbed photons defines the quantum yield of a molecule. Upon excitation, the molecule can relax back to stable ground state via several pathways such as, thermal relaxation, energy transfer, collision, photon emission etc. In general, these processes can be categorised in two classes: radiative and non-radiative. Considering k_r and k_{nr} as the rate constant for radiative and non-radiative processes, the quantum yield is the fraction of fluorophores decaying through emission i.e,

$$\eta = \frac{k_r}{k_r + k_{nr}} \tag{6.7}$$

[9] The Franck–Condon principle is related to the allowed vibrational transitions between two different electronic states and the Franck- Condon factor i.e., $\int \psi_\nu^* \psi_\nu \, d\tau_n$, where ψ_ν is vibronic wavefunction.

where, all the radiative processes are clubbed together, but in practise not all emitted photons can be detected.

In general, the yield is in the range, $0 \leq \eta \leq 1$ that suggests non-radiative losses, but fluorophores can in-principle reach close to 1, especially when non-radiative transitions are minimized. However, photoactivable fluorophores used in single molecule super-resolution microscopy normally have photoconversion yields much less than 1. For example, the photoconversion quantum yield from green-to-red of Dendra2 was estimated to be $(1.4 \pm 0.6) \times 10^{-5}$ under 405 nm illumination [3], whereas, photoconversion quantum-yield of green-to-red photoconvertible fluorescent proteins (monomeric variant of $mKikGR$) is found to be 5×10^{-5} [4].

6.4.2 Fluorescence Lifetime

Once excited, the fluorophore spends some time in the excited state before relaxing to the ground state. This is termed as the lifetime of a fluorophore. Fluorescence lifetime is a key characteristic of a fluorophore since this determines the time available for it to interact with the immediate neighborhood or diffuse or undergo bond formation. In general, the fluorescence lifetime is defined as,

$$\tau = \frac{1}{k_r + k_{nr}} \tag{6.8}$$

Since, lifetime is an average time a molecule spends in the excited state, only a small fraction of molecules emit photons at precisely $t = \tau$. This is due to the random nature of fluorescence emission and the fact that emission probability follows an exponential decay, necessitating 63% and 37% of molecules to decay prior and after τ, respectively.

Another aspect of immense importance is the natural lifetime of a fluorophore, which is the lifetime in absence of non-radiative processes i.e,

$$\tau_n = \frac{1}{k_r} \tag{6.9}$$

Both the interlinked processes (quantum yield and lifetime) are better understood based on Jablonski diagram (a simplified representation of MOT) as shown in Fig. 6.6. The relation between them is given by,

$$\begin{cases} \eta = k_r \times \tau \\ \tau_n = \tau/\eta \end{cases} \tag{6.10}$$

Both fluorescent quantum yield and lifetime along with photoconversion quantum yield and photobleaching quantum yield for few fluorophores used in super-resolution microscopy are briefly given in Table 6.2 [5–10].

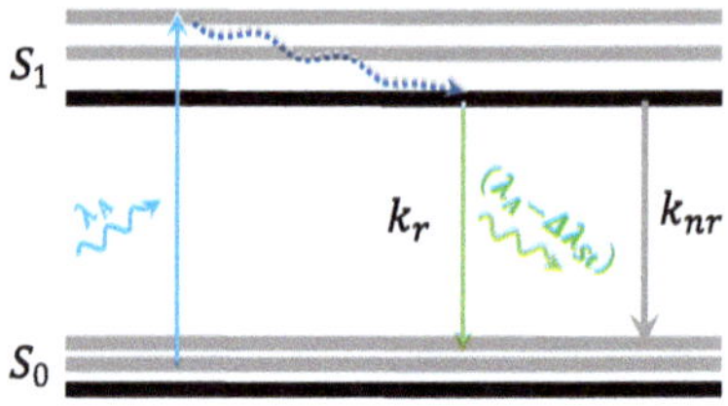

Fig. 6.6 A typical Jablonski diagram illustrating quantum yield and lifetime

Table 6.2 Quantum yield of some well known fluorophores used in fluorescence and super-resolution microscopy. Here, η_F, η_P, η_B and τ represents fluorescence QY, photobleaching yield, blinking-off yield and fluorescence lifetime, respectively [5–10]

	$\lambda_{abs}\|\lambda_{em}$ (nm)	η_F	η_P	η_B	τ (sec.)
paGFP	504|517	0.79	1.1×10^{-6}	$\sim 6 \times 10^{-5}$	2.2 ns
Dendra2	553|573	0.55	2.3×10^{-5}	8.6×10^{-6}	3.3 ns
mEosFP	559|581	0.62	1.6×10^{-5}	3.0×10^{-5}	3.26 ± 0.015 ns
PAmCherry1	564|595	0.46	1.1×10^{-6}	–	–
EGFP	488.5|510.5	0.60	–	–	2.55 ± 0.015 ns
Cy3	647|662	0.18	1.7×10^{-2}	–	0.18 ± 0.0029 ns

6.5 Fluorescent Dyes and Proteins

In this section, we will discuss few fluorophores (both dyes and proteins) that are typically used in fluorescence based techniques and specifically used in single molecule studies.

6.5.1 Fluorescent Dyes

Dyes are long being used as markers for a variety of investigations tools including, fluorescence microscopy, flow cytometry, fluorescence histopathology, and biomedical imaging. In general, fluorescent dyes are conjugated with the protein or organelle of interest to carry out biological/biomedical studies. Accordingly, the dyes can be sub-classified as, exogenous and endogenous. Exogenous are the ones that are introduced in the specimen from outside. Some of the well-known examples of exogenous dyes are, Cyanines, Alexa, Atto, Rhodamines, among others, while endogenous are largely proteins that are either naturally fluorescent (e.g., Tryptophan) or incorporated in the gene via molecular cloning or otherwise. All these dyes will be discussed in the subsequent sections.

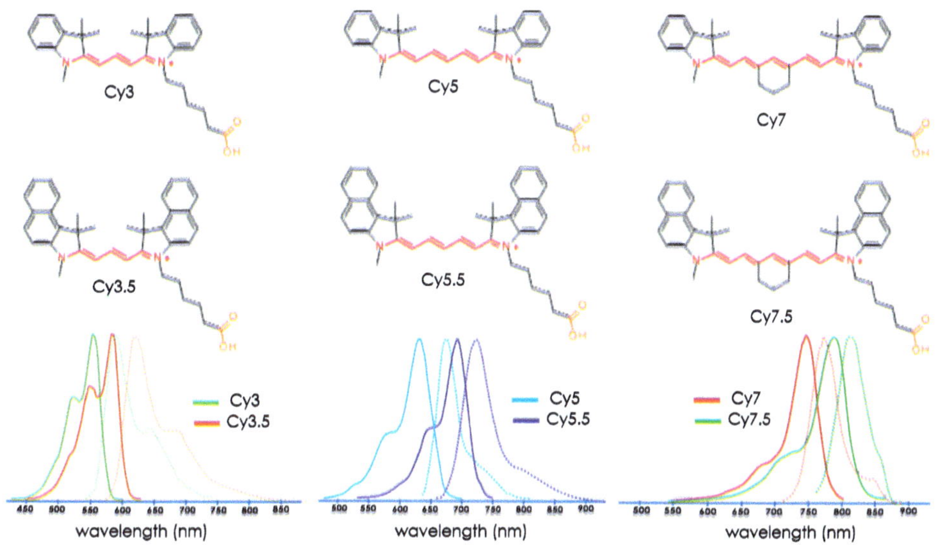

Fig. 6.7 Cynine dyes, derivatives and their chemical structure along with their absorption/emission spectra (Adapted from, https://www.lumiprobe.com/tech/cyanine-dyes)

6.5.2 Fluorescent Dyes for Super-Resolution

Cyanines Also known as tetramethylindo(di)-carbocyanines, these are a class of dyes that are conjugated system between two nitrogen atoms in each resonance structure.

Most of the cyanines and their derivatices have outstandingly high extinction coefficients often exceeding 100,000 mol^{-1} cm^{-1}, which can be largely attributed to their chemical structure.[10] As compared to other dyes, cyanines allow control of key optical properties of the chromophore such as, photoatability, absorption/emission wavelength, upon suitable substituents. Another variety of cyanines (known as, sulfonated-cynines) are also available that allows it to be dissolved in water and are less prone to aggregation. Some of the extensively used non-sulfonated dyes are Cy3, Cy5, Cy7 and their derivatives (fat-solvable $Cy3.5$ and $Cy5.5$). Here the first digit signifies the number of carbon atoms between the indolenine groups, and the decimal suffix (such as, 0.5) represents the benzo-fused cynines. Due to the large variation in the structures, it allows rapid change in fluorescence properties that covers the entire visible spectrum extending to NIR.

Some of the most popular cynanine dyes along with their excitation and emission spectrum are shown in Fig. 6.7. The base chemical structure is that of Cy3 and addition of other functional groups (benzo, nitrogen and others) have changed the absorption spectra covering the entire visible spectrum. These dyes are conjugated with the biomolecule of

[10] Extinction coefficient or molar absorptivity is the measure of a substance's light absorbing ability per unit concentration per unit centimeter path i.e, $M^{-1}cm^{-1}$.

interest (protein, peptide, or DNA) in appropriate aqueous buffer for carrying out specific studies.

Alexa Dyes This family of dyes are obtained from the sulfonation of amino-coumarin or rhodamine molecule. Chemically, this is equivalent to replacing one or more Hydrogen atoms in the dyes with sulfonic acid or its salts. In addition, the introduction of sulfonic group gives a net negative charge on the dye that makes it water soluble or more hydrophilic. Moreover, sulfonation efficiently scales the excitation wavelengths from UV (350 nm) to orange (594 nm) [11].

Studies show that Alexa family of dyes have strong fluorescence and enhanced brightness when compared to other dyes. Even though some of the the dyes (e.g. Alexa 350 and Alexa 430) have low extinction coefficient ($<20{,}000\,\mathrm{cm^{-1}M^{-1}}$), but are preferred for imaging which is due to their large Stokes shifts (extending beyond 70 nm). In addition, lower molecular weight and photostability adds to its advantage as an efficient fluorescent probe. These qualities makes them perfect probe-of-choice for bioimaging. Studies over the last two decades have established successful conjugation of Alexa dyes with proteins, antibodies and organelle specific biomolecules. A few Alexa dyes along with their excitation and emission spectra is shown in Fig. 6.8.

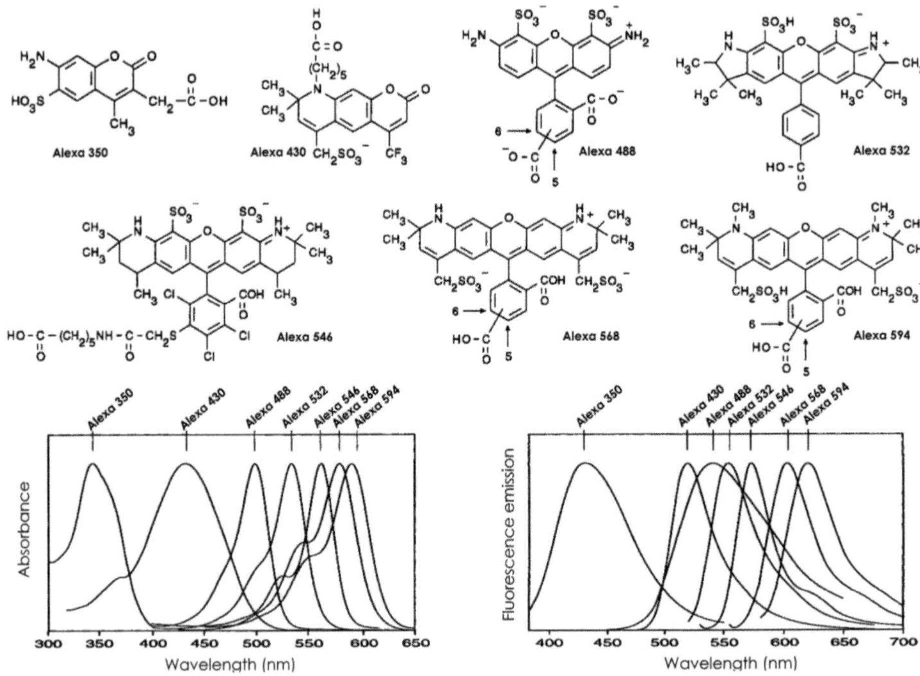

Fig. 6.8 Some of the popular Alexa dyes (Alexa 350–594) along with their absorption/emission spectra (Adapted from N. Panchuk-Voloshina et al., Jl. Histochem. Cytochem. 47, 1179–1188 (1999))

Other fluorescent probes such as, Bodipy, Atto, Rhodamine, TAMRA and their derivatives are routinely used for a variety of fluorescence based assays (from microscopy to histopathology).

6.5.3 Proteins for Fluorescence Imaging

The next set of fluorescent probe belongs to endogeneous class and are largely proteins that can fluoresce and have properties just like dyes. These probes have the advantage of strong fluorescence upon activation, easy incorporation in biological processes via genetic fusion, and a precision reporter for a protein-of-interest. Here we discuss few class of fluorescent proteins that are frequently used in fluorescence and super-resolution microscopy.

GFP and Derivatives Green fluorescent protein is the first fluorescent proteins ever discovered. Osamu Shimomura was first to isolate GFP from the jellyfish *Aequorea Victoria* in the year 1961. Together with Martin Chalfie and Roger Y. Tsien, he received Nobel prize in 2008 for the discovery and development of the green fluorescent protein, GFP. Most of the GFP and its variants are around 730 base-pairs long, has about 238 amino acids and has a molecular mass of 27 kD. GFP obtained from A. victoria has excitation peak at 395 nm and 475 nm, whereas GFP from sea pansy (*Renilla Reniformis*) has a peak at 498 nm, and the emission is predominantly at 509 nm. Wild-type $Aequoreavictoria$ GFP has a quantum yield of 0.79. Figure 6.9 shows GFP and some of its spectral-derivatives along with their characteristic fluorescent spectrum. Tailored shifts in wavelength engineered by fluorophore and β-barrel mutations help provideoptions across the entire visible spectrum. This provides a convenient choice for single and multi-color imaging.

6.5.4 Fluorescent Proteins for Super-Resolution

Recent advances in super-resolution microscopy has seen a number of photoactivatable fluorescent proteins ($paFPs$) capable of a multifold increase in fluorescence brightness when subjected to a specific wavelength of light and intensity. Some of the key properties of a fluorescent protein are, strong fluorescence, monomeric in nature, relatively low dark state occupancy after photoactivation, fast maturation and high photostability [12]. Specifically, the monomeric nature and maturation are important for it to be used as a fusion protein and early detection. pH stability is another important factor, especially for live cell imaging.

These class of proteins needs a separate mechanism for it to be used for super-resolution imaging. The properties of single molecule proteins are harnessed to enable sparse blinking that enables determination of its position and localization in the specimen.

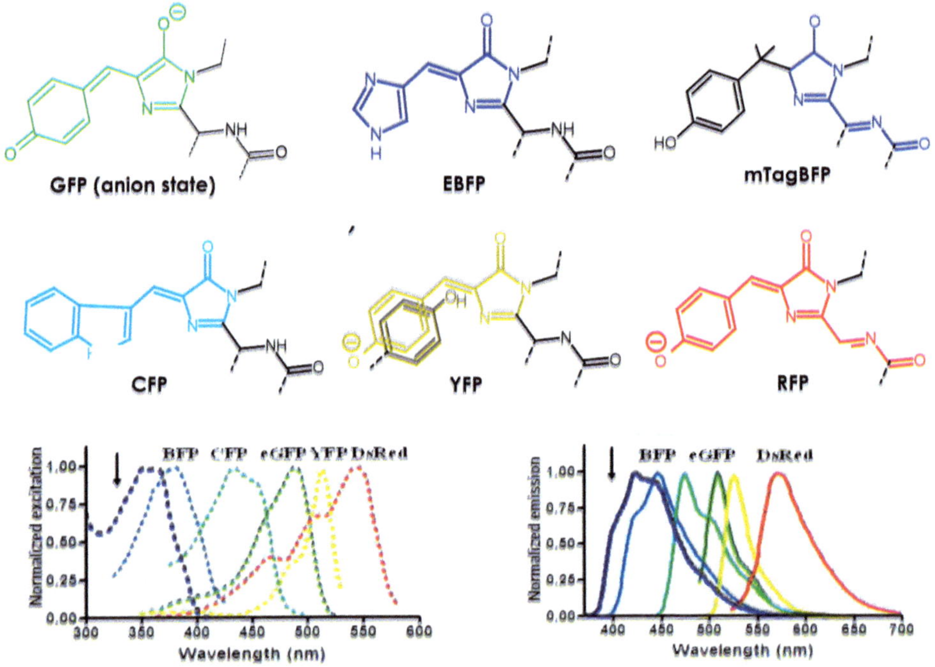

Fig. 6.9 Green fluorescent protein (GFP) and its derivatives along with their absorption/emission spectra (Adapted from, PATENT, Fluorescent protein with deep blue color, KR20100065294A South Korea (2010))

Specifically, this class of molecules needs activation before producing a measurable fluorescence signal. In general, activation requires relatively high energy transfer to the molecule to change its state from dark state or weak fluorescence to strong fluorescence with shift in wavelength. However, this is not a strict requirement, and activation can be achieved by other means. There are several pathways to affect activation which we will be discuss in Chap. 8. Here, we will concentrate on some of the popular photoactivable proteins used in super-resolution microscopy.

paGFP This probe is among the very first photoactivable fluorophores which is a photoactivable variant of Aequorea victoria green fluorescent protein (GFP). It is derived from $wtGFP$ by the substitution of histidine to threonine at position 203 [13]. paGFP emits weak fluorescence, however upon activation with a 405 nm light, the fluorescence increases by many folds in green with its emission maximum at 504 nm. This gives it advantage over general fluorophore since this produces high contrast during imaging. In general, paGFP is bright, photostable and is a monomer i.e, compatible for fusion protein, thereby making it a good choice for biological study.

Dendra2 This is an improved commercially available version of a convertible protein (green-to-red) Dendra. Dendra2 is monomeric in nature allowing protein labeling, has fast maturation (chromophore formation) at 37°C in mammalian cells, high contrast photoconversion, high photostability of the activated red signal, low phototoxic activation at 488 nm, and is bright both before & after photoswitching. Dendra2 (non-activated) has a normal excitation at 490 nm with fluorescence peaked at 507 nm. Post-activation with a 405 nm or 488 nm light, the absorption shifts to 553 nm with strong emission at 573 nm [14]. These properties make Dendra2, a monomeric green-to-red photoconvertible fluorescent protein for live cell imaging and protein tracking in living cells.

PamCherry1 The protein is derived from DsRed which switch from dark to red upon irradiation with violet 399 nm light. As compared to paGFP, they are less phototoxic and hence suitable for depth imaging. PAmCherry1 is a monomer and hence are suitable for fusion tagging other proteins of interest. Before photoactivation, the PAmCherry1 has a absorbance maxima at 405, but does not fluoresce well when excited at these wavelengths. However, when activated, PAmCherry1 exhibited red fluorescence with excitation/emission peaks at 564/595 [15]. Since this has a higher quantum yield, better contrast ratio, and fast maturation rate, it is preferred as a monomeric red photoactivable fluorescent protein. The quantum yield is quite decent (0.24–0.53), and maturation half time is 1.2–1.7- fold to that of $mCherry$ at 37°C. Note that, PAmCherry1 together with paGFP is a great combination for two-color super-resolution imaging [16].

mEos This is a monomer derived from Eos [17]. It is bright, and has a good contrast in the green-to-red highlighters. It has the highest photon output among the green-to-red photoconvertible fluorophores, and is compatible as a fusion protein. Hence, with paGFP this is a formadible combination for multicolor imaging. $mEos$ is stable over a large temperature range, highly fluorescent, and has a good photostability, making it a better choice among other green-to-red photoconverters. $mEos$ (green form) is weakly excitable at 390 nm for which the emission can be observed at 540 nm. However upon irradiation at 390 nm causes conversion of the green-to-red emitting state i.e, the spectrum shifts to red form which is excitable at 560 nm with an emission at 600 nm.

Further details related to the structure and excitation/emission spectra of the above photoactivable probes for both pre- and post-activation is shown in Fig. 6.10.

References

1. Stokes GG (1852) On the change of refrangibility of light. Phil Trans R Soc (London) 142:463–562

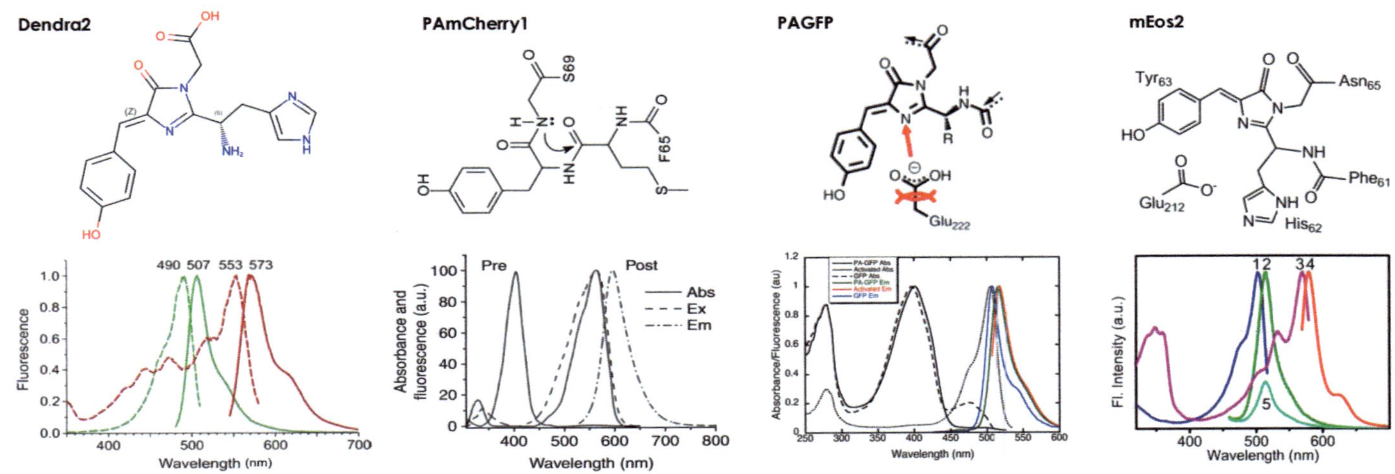

Fig. 6.10 Chemical structure of few non-activated forms of photoactivable fluorescent molecules (Dendra2, PAmCherry1, PAGFP and mEos2). The corresponding molecular spectra for both activated and non-activated forms are also shown [12–17]

2. Kasha M (1950) Characterization of electronic transitions in complex molecules. Disc Faraday Soc 9:14–19
3. Avilov S et al (2014) In cellulo Evaluation of Phototransformation Quantum yields in fluorescent proteins used as markers for single-molecule localization microscopy. PLoS One 9(6):e98362
4. Habuchi S, Tsutsui H, Kochaniak AB, Miyawaki A, van Oijen AM (2008) mKikGR, a monomeric photoswitchable fluorescent protein. PLoS One 3:e3944
5. Thompson MA, Biteen JS, Lord SJ, Conley NR, Moerner WE (2010) Molecules and methods for super-resolution imaging. Methods Enzymol 475:27–59
6. Testa I et al (2006) T2P-GFP: two-photon photo-activation of PA-GFP in the 720–840 nm spectral region, In: Proceedings of the SPIE 6089, Multiphoton Microscopy in the Biomedical Sciences VI, p 608912
7. Osuga M, Nishimura T, Suetsugu S (2021) Development of a green reversibly photoswitchable variant of Eos fluorescent protein with fixation resistance. Mol Biol Cell 32(21):br7
8. Kang J (2020) Restricted intramolecular rotation of fluorescent molecular rotors at the periphery of aqueous microdroplets in oil. Sci Rep 10:16859
9. Fron E et al (2013) Revealing the excited-state dynamics of the fluorescent protein Dendra2. Jl Phys Chem B 117:2300–2313
10. Avilov S et al (2014) In cellulo Evaluation of Phototransformation quantum yields in fluorescent proteins used as markers for single-molecule localization microscopy. PloS One 9:e98362
11. Panchuk-Voloshina N et al (1999) Jl. Histochem. Cytochem 47:1179–1188
12. Pennacchietti F, Gould TJ, Hess ST (2017) The role of probe photophysics in localization-based superresolution microscopy. Biophys J 113:2037–2054
13. Patterson GH, Lippincott-Schwartz J (2002) A photoactivatable GFP for selective photolabeling of proteins and cells. Science 297:1873–1877
14. Chudakov DM, Lukyanov S, Lukyanov KA (2007) Using photoactivatable fluorescent protein Dendra2 to track protein movement. BioTechniques **42**:553–563
15. Subach FV (2009) Photoactivatable mCherry for high-resolution two-color fluorescence microscopy. Nat Meth 6:153–159
16. Lippincott-Schwartz J, Patterson GH (2009) Photoactivatable fluorescent proteins for diffraction-limited and super-resolution imaging. Trends Cell Biol 19:555–565
17. Saha R et al (2013) Light driven ultrafast electron transfer in oxidative redding of Green Fluorescent Proteins. Sci Rep 3, 1580

7 The Photophysics/Photochemistry of Single Molecule Probes

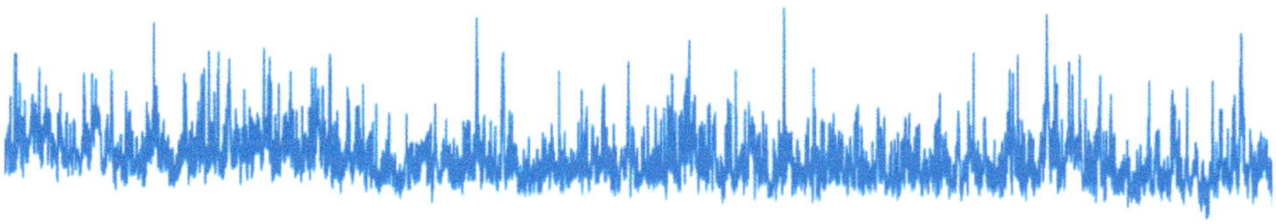

The probes used in fluorescence-based assays/imaging have the distinct advantage of reducing non-specific signals and bypassing scattering, giving them an edge over other imaging techniques. At the core of this lies, the fluorophores (proteins and dyes) with unique chemistry and unique mechanisms. This is specifically true with fluorophores that are photoactivable/photoconvertible/photoswitchable. Molecular cloning and conjugation methods allow the tethering of these probes with biomolecules and proteins of interest and their targeted delivery to the organelle of activity. The conversion of one to another form occurs through a photochemical process whereby small chemical groups is removed or modified. For example, Cy5 converts to Cy3 via novel excision of C_2H_2 from the polymethine chain. Another mechanism for photoconversion involves the conformation of arginine at residue 66 by threonine at residue 69 of fluorescent proteins from Anthozoan families (Dendra2, maple, Eos, mKikGR, pcDronpa protein families). Some fluorescent proteins like wt-GFP exist in two forms: neutral and anionic, of which one form (mostly anionic) is fluorescent. These fluorescent proteins often link through hydrogen bonds to other proteins of interest/biological macromolecules, making it the most desired probe for live cell/organism imaging. In some photoswitchable proteins (IrisFP and Dronpa), a unique mechanism takes place where Cis-Trans isomerization of the chromophore results in photoconversion. In this chapter, we will focus on the photochemistry of these unique fluorescent probes and discuss their mechanism of action, which gives them central stage in super-resolution imaging.

P. P. Mondal, S. Hess, *Single-Molecule Super-Resolution Optical Microscopy*,
https://doi.org/10.1007/978-3-032-13442-4_7

The last decade has seen a tremendous growth in the development of single molecule probes. This can be largely attributed to the development in photochemistry of fluorescent probes. In general, the probes used in super-resolution microscopy can be classified based on their chemical nature: exogenous probes which are inherently dye based, and endogenous probes that are protein-based.

Dye based exogenous probes (such as, Cyanines, Alexa Fluor etc) are preferred to visualize sub-cellular organelles. This is due to their high quantum efficiency, brightness, and stability. Often these dyes are chemically conjugated to a specific biomolecule of interest for visualizing them in the targeted organelles. Dye-conjugation works by chemically linking or attaching a fluorescent dye to the molecule of interest or antibody. This refers to a covalent bond between them, creating a new hybrid molecule that inherit the property of both dye and target molecule. For example, antibodies conjugated with dyes like Cy3, Cy5, Cy5.5 or Alexa Fluor are used for following specific proteins in cells or organisms. DNA probes labeled with dyes are used for identifying specific DNA sequences, and drugs are conjugated to dyes for targeted delivery to tumor cells. Several such studies are reported where exogeneous probes are used and have shown future potential.

The other class of fluorescent probes are protein based endogenous probes (such as, GFP, PAGFP, Dendra2, mEos3.2 etc) which are preferred for studies that require prolonged observation of biological processes in a live specimen (cell/organism). These probes can be internalized in a cell at DNA level that allows its expression and production in the cellular environment. However, these probes are not as bright as that of Dye probes and are often unstable, leading to photobleaching. The protein conjugation for fluorescent proteins is achieved using molecular cloning that allows the desired protein fusion into a cloning vector. This involves designing and inserting specific DNA sequences into a cloning vector. In short, the process involves, (1) DNA isolation (obtaining a source DNA using restriction enzymes or PCR), (2) vector selection (plasmid containing elements of replication, selection and protein expression), (3) Ligation (joining gene-of-interest and vector), constructing a recombinant DNA molecule, (4) Transformation (recombinant vector induced into a bacteria), (5) Amplification and Expression (the bacteria is grown and gene-of-interest is expressed), (6) Purification (extracting the plasmid). All these process are performed sequentially to obtain the target DNA plasmid with a good concentration ($\sim 10^3$ ng/μlit) and a healthy DNA purity absorbance ratio (260/280 $\sim$ 1.8 and 260/230 $\sim$ 2–2.2). The plasmid is introduced to a host cell and the cell machinery process the plasmid DNA to produce the protein-of-interest. This is used in a number of applications including, DNA sequencing encoding, targeted drug delivery and disease progression studies. Detailed discussion can be found in Chap. 10.

In this chapter, we will discuss both types of fluorescent probes, their mechanism of action specific to super-resolution imaging, and give examples of few probes.

7.1 General Photochemistry of Molecular Probes

The mechanism by which single molecules can toggle between bright and dark states forms the basis of single molecule imaging. Specifically the molecules that are capable of switching phenomena (either once or many times) are key to super-resolution microscopy.

To understand the mechanism, let us consider a simple molecule with the electronic states as shown in Fig. 7.1. When fluorophores are excited from the ground state S_0, it can relax to the ground state via multitude of processes as shown in Fig. 7.1. The immediate process is the internal conversion where the molecule relaxes to the lowest vibrational level of first excited singlet state S_1. Subsequently, the fluorophore can go back to the ground state S_0 via radiative and nonradiative pathways or makes a transition to the energetically favorable lowest triplet state by intersystem crossing (ISC). Since the triplet state is a metastable state and have a relatively longer lifetime ($10^{-6} - 10^{-3}$ s), it is prone to a several processes including reaction with reactive oxygen species. Often, reaction with reactive triplet oxygen to produce singlet oxygen ($^1O_2$). Another possibility is the fluorophore undergoing redox reactions to produce radical cations (R^+) or anions (R^-). These processes may result in steady fluorescence ($S_1 \leftrightarrow S_0$) or blinking ($S_1 \rightarrow T_1 \rightarrow S_0$) or fluorescence followed by permanent bleaching ($T_1 \rightarrow R^+,\ R^-$ or $^3O_2 \rightarrow^1 O_2$) as shown in Fig. 7.1.

These possibilities lead to the ultimate decomposition of chemical structure, resulting in a transformation from fluorescent to permanent non-fluorescent form i.e, inhibit fluorophores ability to fluoresce (photobleaching).

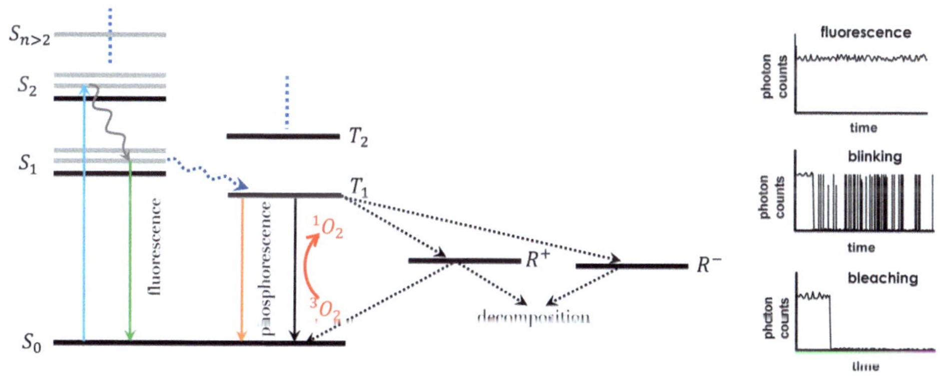

Fig. 7.1 Photochemical processes of an excited fluorophore. Right panel shows the signature of fluorescence, blinking and photobleaching

7.2 Single Molecules Switching

We will dedicate this section to single molecules that blink and also discuss the underlying blinking mechanism. In general, fluorophores that enable single molecule imaging are of two major types: exogenous synthetic dyes and endogenous fluorescent proteins. Within these types, there are four different classes: photoswitchable, photoactivable, photoconvertible and spontaneously blinking dyes. While synthetic dyes are bright and are better localized enabling high resolution, fluorescent proteins are suited for live-cell imaging. This makes fluorescent proteins more potent for disease biology and drug discovery [1–3].

It may be noted that, most of the fluorophores used in single molecule imaging switch between ON and OFF states, but the exact molecular mechanism is still unknown for a large number of probes. In this section, we will investigate some of the well-known probes used in single molecule imaging and give few known mechanisms.

7.2.1 Photoswitchable Fluorophores

These specialized fluorophores are known to undergo reversible transitions between ON and OFF states that allow them to blink several times before actual photobleaching. Some of the well-known photoswitchable dyes include cynine dyes, Alexa Fluor dyes, and ATTO dyes [4]. The switching between ON and OFF states can be carried out using photo-switching buffer (PBS) with thiol (mercaptoethylamine) or upon irradiation by an UV/blue light. A general switching mechanism of a typical photoswitchable fluorophore and the associated photochemical processes is shown Fig. 7.2 [5]. When excited by an appropriate wavelength of light, the molecule makes a transition from ground to first excited Singlet state. Once in excited state, the molecule can either follow the fluorescence cycle or non-fluorescence pathways for de-excitation with a certain probability. While the molecule initially follows fluorescence cycle giving out fluorescence, there is a finite chance that the

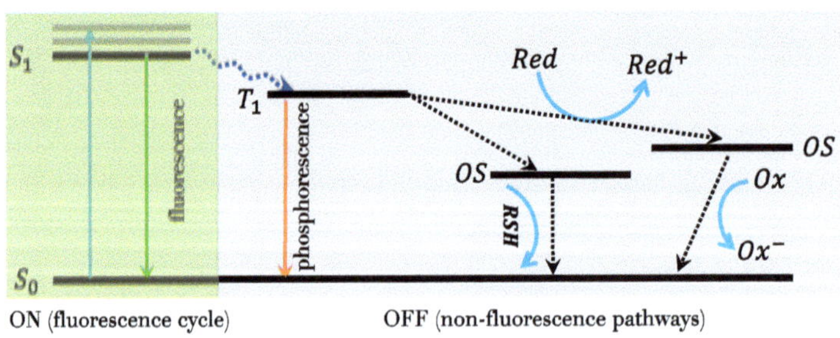

Fig. 7.2 Photoswitching of fluorophores as a result of switching between ON states ($S_0 \leftrightarrow S_1$) and OFF states (T_1, other long-lived states, OS). Addition of reductants (Red), thiols (RSH) and oxidants (Ox) help rapid recovery of fluorophores before permanent photobleaching

molecule makes a transition to triplet which is a metastable state (longer lifetime), giving opportunity to chemically react with other species (thiols RSH, Oxygen or reductants). This produces non-fluorescent radical anions with a lifetime of tens of milliseconds, thereby switching-off the fluorescence [6,7]. Moreover, these dark non-fluorescence states are also reduced by buffer thiol to produce reduced OFF states which are stable for tens of milliseconds. Subsequently, the fluorophore returns to the ground state (S_0) upon reaction with oxygen. Note that, direct irradiation of OFF-state molecule with UV can also be used to increase activation of fluorophores. In general, the ON-state has a typical lifetime of 5–30 ms, although it can be as short as 1 ms and as large as 500 ms. [8–10]

7.2.2 Photoactivable Fluorophores

Photoactivable fluorophores are a distinct class of molecules that switch from OFF to ON state either spontaneously or upon activation by UV light (mostly, 405 nm). In general, photoactivation are of two types: reversible photoswitching between fluorescent and nonfluorescent forms and irreversible photoconversion [11].

Figure 7.3 shows photoactivation mechanism of a typical fluorophore (IrisFP). Note that IrisFP is a photoactivatable fluorescent protein that combines irreversible photoconversion from a green- to a red-emitting form with reversible photoswitching between a fluorescent and a nonfluorescent state in both forms [12]. Upon UV light excitation, the protein chromophore undergoes the excited state trans-to-cis isomerization on a picosecond timescale, wheread the deprotonation occurs on the micro-to-milli second timescales. Specifically, decay can occur from multiple closely lying excited states (such as, $I1^*$, $I2^*$) depending upon the nature of decay (sequential or parallel) [13]. Finally, the relaxation to the ground state occurs either via green trans-protonated state (OFF state) or via green cis-

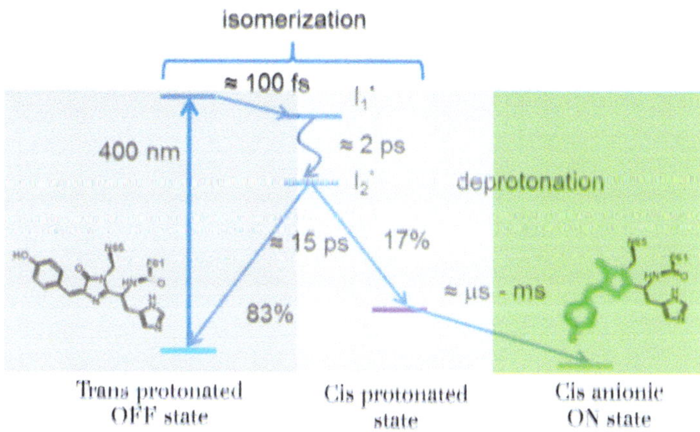

Fig. 7.3 Photoactivation of a typical fluorophore as a result of transition between Cis-Trans states. The figure is an adaptation from J P Colletier et al., Jl. Phys. Chem. Lett. 7, 882–887 (2016)

protonated state (ON state). So, a fast change in the structural dynamics of a photoactivable fluorophores gives rise to blinking. Some of the fluorophores that closely follow this mechanism are, *IrisFP* and *Dronpa* [14].

From imaging point of view, the fluorophores are initially OFF, they are activated with UV light. For localizing single molecules, photobleaching by a light of longer wavelength (usually excitation wavelength) is carried out, prior to activation of new fluorophores. The single molecule PSFs are then recorded for a prolonged time to enable complete image formation of the target subcellular structure. Usually fluorophores with long ON-state lifetimes and higher photon yields are preferred. This include, photoactivable dyes such as Janelia Fluor dyes and rhodamine PA Janelia Fluor [15]. However, for live cell imaging it is the photoactivable proteins that play a major role [10]. Some of the widely-used photoactivable proteins include, PA-GFP ($\lambda_{OFF \rightarrow ON} = 405$ nm, $\lambda_{ON \rightarrow OFF} = 488$ nm), PAmCherry ($\lambda_{OFF \rightarrow ON} = 405$ nm, $\lambda_{ON \rightarrow OFF} = 564$ nm), PAmKate ($\lambda_{OFF \rightarrow ON} =$ 405 nm, $\lambda_{ON \rightarrow OFF} = 586$ nm) [9, 16].

7.2.3 Photoconvertable Fluorophores

This class of fluorophores switch irreversibly from one spectral state to another. Among others fluorophores, green to red photoconvertible fluorophores (pnFPs) change their fluorescence spectra from green to red type fluorescence upon irradiation by a high energy light (ultraviolet) irreversibly. These fluorophores hold great potential primarily due to their brightness, photostability and photoconversion contrast, among others. Some of the well-known fluorophores are, Keade, Eos, Dendra2, mKikGR and mMaple [17–19].

Figure 7.4 shows a general mechanism of photoconversion for a large fraction of nonconvertible fluorophores. Initially, the fluorophore stays in the green form. Both the

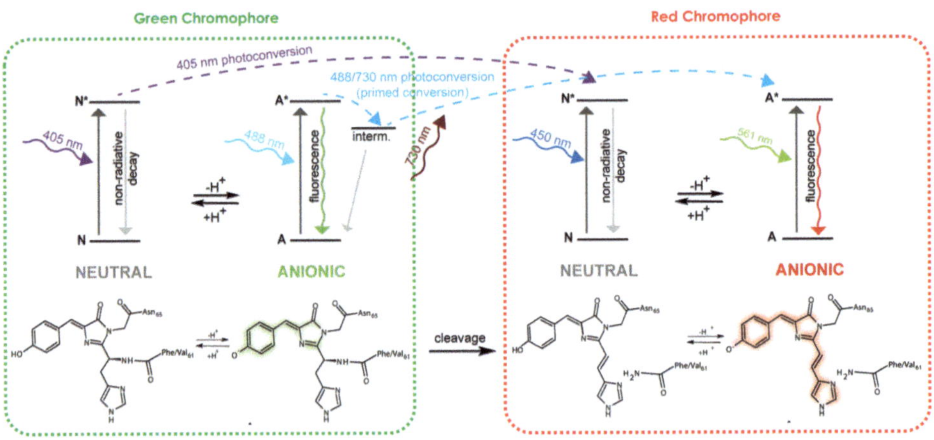

Fig. 7.4 Photoconversion mechanism of a typical photoconvertible fluorophore. The figure is an adaptation from B. Turkowyd et al., Angew. Chem. (Int. Ed.) 56, 11634–11639 (2017)

neutral and anionic state are in equilibrium. Upon irradiation with 488 nm light, only the excited (A^*) anionic chromophore fluoresce in the 500–520 range. A light-induced cleavage of the fluorophore backbone occurs in its anionic green form. Note that, the red chromophore exist in both neutral and anionic states and are in equilibrium. However, the anionic state of the red form fluoresces upon direct irradiation with a 561 nm light. The chromophore also photoconverts from its anionic green upon irradiation with 488 nm light. Dendra2 is a good example of this. The green anionic chromophore is primed into a millisecond long intermediate state (triplet or other states) from which the conversion to red state can be induced by 730 nm light. This suggests a new switching mechanism (primed conversion) for Dendra2 using combined illumination at 488 nm and 730 nm [20, 21]. Although certain aspects are understood but a complete understanding of the underlying mechanism is yet to be done. Overall, photoconvertible fluorophores (proteins) are good choices for imaging live biological specimens.

There are dyes (such as, HMSiR and FRD) that are known to blink spontaneously. The advantage is that a separate light or buffer is not required for photoswitching. These dyes show immediate blinking and have ON-state lifetimes in the range of 10–300 ms [22]. These dyes have excellent photon yield and are membrane permeable, making them useful for live cell imaging.

7.3 Fluorophores/Dyes/Proteins

In this section, we will discuss a range of fluorophores (both dyes and proteins) and their applications.

7.3.1 GFP and Derivative Proteins

GFP is one of the earliest known fluorescent proteins that has brought revolution in fluorescence microscopy [23–25]. The protein is extracted from the gellyfish *Aequorea victoria*, and has an atomic mass of 27 kDa. The chromophore of this protein has 238 amino acid residues and is embedded in the protein matrix, isolating it from the bulk. The chromophore of wild-type GFP has two absorption peaks one at 395 nm and other at 475 nm which are attributed to the neutral and anionic forms. However, there is a single emission peak at 508 nm with a high fluorescence quantum yield of 0.79 [26, 27]. The structure of GFP, the excitation mechanism and the absorption/emission spectra is shown in Fig. 7.5. The structure of GFP folding into 11-strand β-barrel is the most likely reason for its bioluminescence. The β-can structure protects the chromophore and gives rise to its stability. The photoisomerization of wild-type GFP necessitates that the neutral form (A-form) goes to the intermediate form (I-form) before converting to the anionic species (B form). Note that, change from form A to I is a protonation change while I to B is a conformational change (through side-chain confirmation of Thr203) [28].

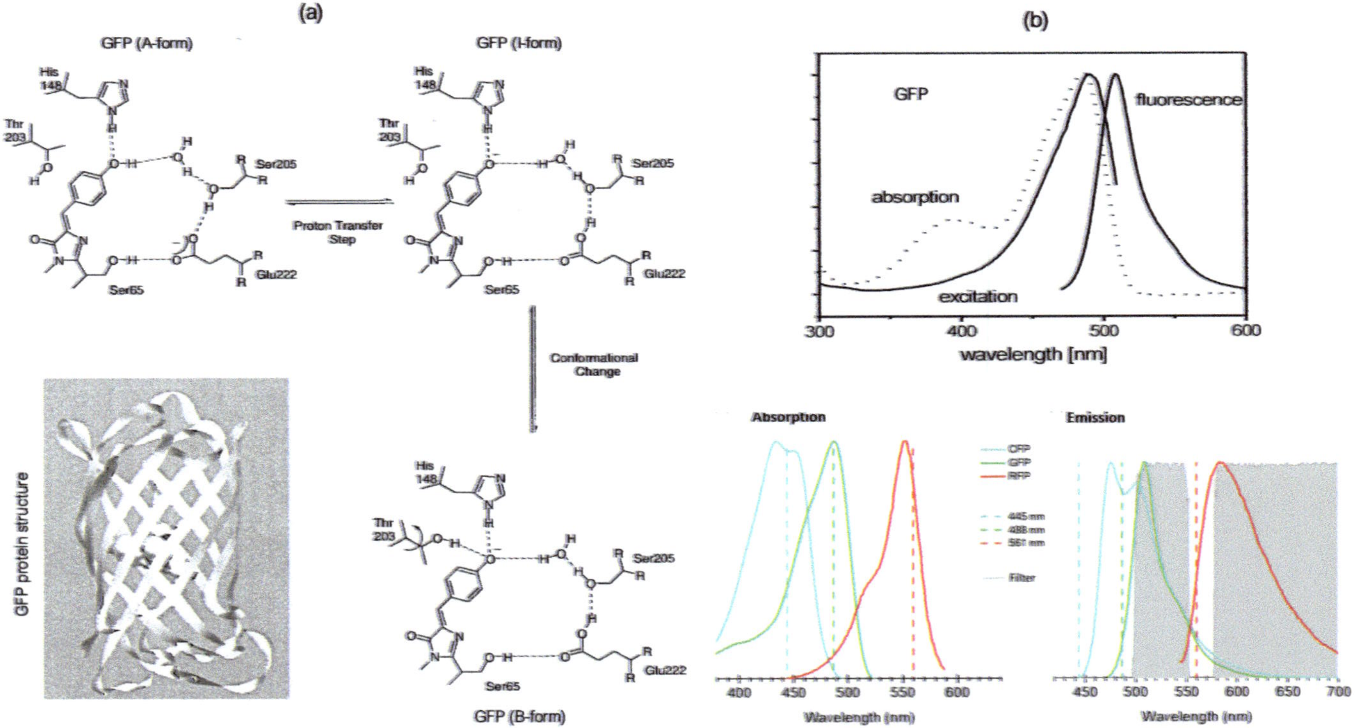

Fig. 7.5 (**a**) GFP structure, along with the mechanism for the photoisomerization of wild-type GFP that shows that the neutral form of the chromophore A convert to the anionic form B by going through the intermediate state I. The change from A to I is a protonation change, however the change from I to B is a conformational change. (**b**) The excitation and emission spectra of GFP and its derivatives (CFP and RFP). The figure is an adaptation from, Chem. Rev. 2002, 102, 759–781; Korean Chem. Soc. 26, 413–417 (2005); Jl. Biomed. Optics 24, 016501 (2019)

The corresponding excitation and emission spectrum of wild-type GFP (wtGFP) is shown Fig. 7.5b. This clearly indicates two peaks in the absorption spectra, one at 395 nm and a dominant peak at 475 nm. The band at 395 nm is attributed to protonated chromophore which occurs due to denaturation or pH change. However, other variants such as CFP and RFP are formed by change in chromophore and have distinct absorption/emission peaks as shown in Fig. 7.5b. The dashed lines indicate laser lines that can be used to efficiently excite CFP/RFP, and the shaded area represents typical multiband filters. This suggests the diverse nature of GFP and its variants for performing multicolor imaging of specimens allowing simultaneous visualization of multiple organelles and their dynamics.

7.3.2 PAGFP, PSCFP2 and Derivatives

PA-GFP and PS-CFP2 are the Aequorea-derived mutants of GFP-like protein from the jellyfish Aequorea coerulescens. These proteins undergo photoactivation by UV-light that induce decarboxylation of the glutamate residue (Glu222), transiting it from neutral to anionic state. Figure 7.6a shows the photoconversion mechanism of PS-CFP2 [29]. Before photoconversion, the chromophore is in a neutral (protonated) state, which fluoresces in the cyan region with its peak at 468 nm. However, this changes drastically after photoconversion by irradiation with a 400 nm light. This causes the decarboxylation of Glu222 residue where rearrangement of H-bonds occur, leading to a transition of the chromophore into an anionic (deprotonated) state. This state emits in the green region with its maximum at 511 nm when excited with 488 nm light. The spectral properties of $PA - CFP2$ before and after photoconversion is shown in Fig. 7.6b.

7.3.3 Dendra and Dendra2 Proteins

Dendra, Dendra2 and related derivative proteins are monomeric and mature to green fluorescent state [29]. In general, the chromophore of Dendra2 exists in an equilibrium between two states: neutral (nonfluorescent) and anionic (green fluorescent) state. After photoconversion using a light of wavelength 405 nm, it photoconverts to a very efficient red state. This change is due to the appearance of double bond ($c\alpha = c\beta$) in the chromophore, forming His residue and the cleavage of protein backbone. Alternately, the green anionic form of Dendra2 can also be converted to red state by irradiation at 490 nm. As compared to 405 nm induced photoconversion, this process is relatively less efficient. This is indicated in Fig. 7.7a by the emission color.

The corresponding absorption and emission spectra of Dendra2 is shown in Fig. 7.7b. Before photoconversion, the protein has two prominent peaks: 386 nm and 488 nm, for which the emission peaks at 507 nm. After photoconversion, the absorption spectra shifts from blue to the green region with peak at 553 nm for which the emission occurs at 573 nm (maximum). The irradiation with 400 nm light cause an irreversible transition into

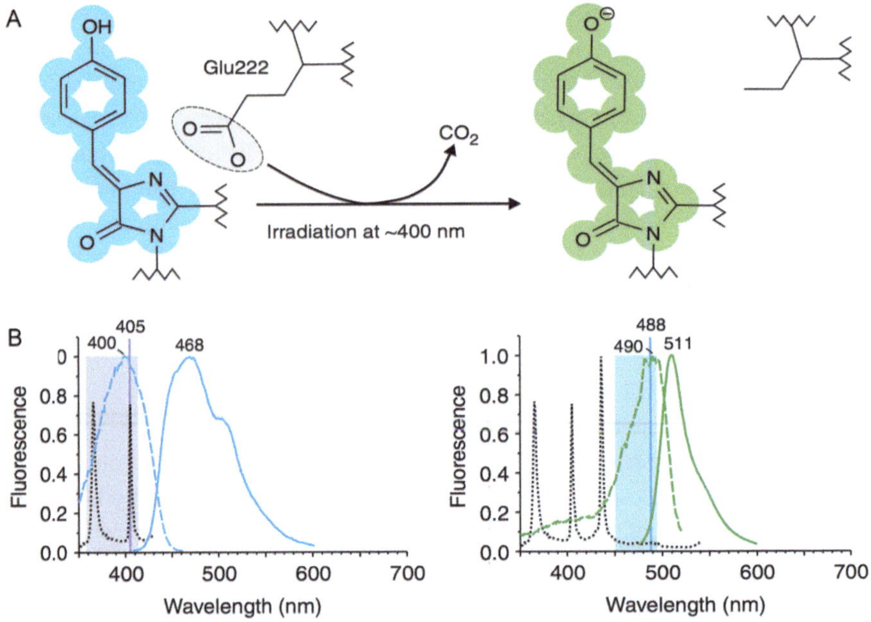

Fig. 7.6 (**a**) Chemical representation of PS-CFP2, along with the photoisomerization mechanism showing conversion from neutral to fluorescent form. (**b**) The excitation and emission spectra of PS-CFP2. Dashed lines: excitation spectra, solid lines: emission spectra. The figure is an adaptation from, Dmitriy M Chudakov et al., Tracking intracellular protein movements using photoswitchable fluorescent proteins PS-CFP2 and Dendra2, Nature Protoc. 2, 8 (2007)

red fluorescent state. Key details related to both PS-CFP2 and Dendra2 is tabulated in Table 7.1.

7.3.4 Cynanine Dyes

Dyes are among the brightest fluorophore that facilitate single molecule imaging with a high quantum yield and large contrast. There are a number of advantages over fluorescent proteins. Among others, Cyanines dyes (Cy3, Cy5, Cy7 and Alexa Fluor 647 (AF647)) are the most extensively used organic fluorophores for single molecule imaging and super-resolution microscopy. This is due to their high molar extinction coefficients, tunability of fluorescence wavelength and targeted labeling [30].

Figure 7.8 shows few popular cyanine dyes and their photoconversion mechanism. Upon irradiation with an excitation light of 650 nm, the photoconversion of Cy5 to Cy3 occurs due to the generation of singlet oxygen which mediate oxidative cleavage of polymethine chain. The resulting carbonyl compound undergoes two- and one- carbon cleavage reactions (see, carbonyl compounds, 1a,b and 2a,b in Fig. 7.8a) to give 3a,b. This leads to the formation of $pcCy5 - ac$ (i.e., $Cy3 - ac$) due to the condensation of 2a,b

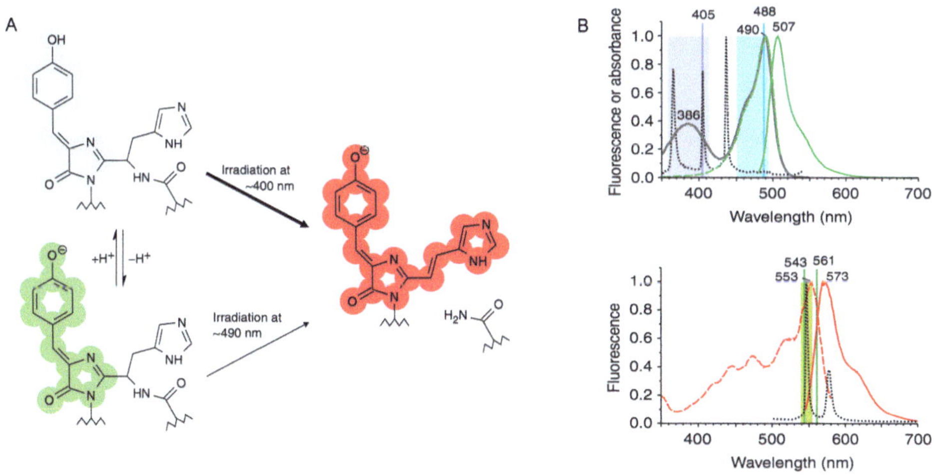

Fig. 7.7 (**a**) Chemical representation of Dendra2, along with the photoisomerization mechanism showing conversion from neutral to fluorescent form. (**b**) The excitation and emission spectra of Dendra2. The figure is an adaptation from, Dmitriy M Chudakov et al., Tracking intracellular protein movements using photoswitchable fluorescent proteins PS-CFP2 and Dendra2, Nature Protoc. 2, 8 (2007)

Table 7.1 Characteristics of PS-CFP2 and Dendra2. The table is adapted from Ref. *Chudakov et al., Nature Protoc. 2, 8 (2007)*

	PS-CFP2		Dendra2	
	Before activation	After activation	Before activation	After activation
Excitation	400 nm	490 nm	490 nm	553 nm
Emission	468 nm	511 nm	507 nm	573 nm
Color	Cyan	Green	Green	Red
Quantum Yield	0.2	0.23	0.50	0.55
Extinction Coeff. ($M^{-1}cm^{-1}$)	43,000	47,000	45,000	35,000
Brightness	0.26	0.33	0.68	0.58
Structure	Monomer		Monomer	
Contrast	> 2000		> 4000	

(Fischer's aldehyde) with 3a,b (Fischer's base). Overall, the excision of C_2H_2 arises from the polymethine chain.

A similar photoconversion chemical process occurs for other cyanine dyes, Cy5.5 and AF647 which is due to photoinduced cleavage of two-carbon units. Similar to Cy5 photoconversion, the excision of C2H2 also occurs for these dyes as shown in Fig. 7.8b,c. Additional details of the photoconversion process can be found in Ref. [31] and associated

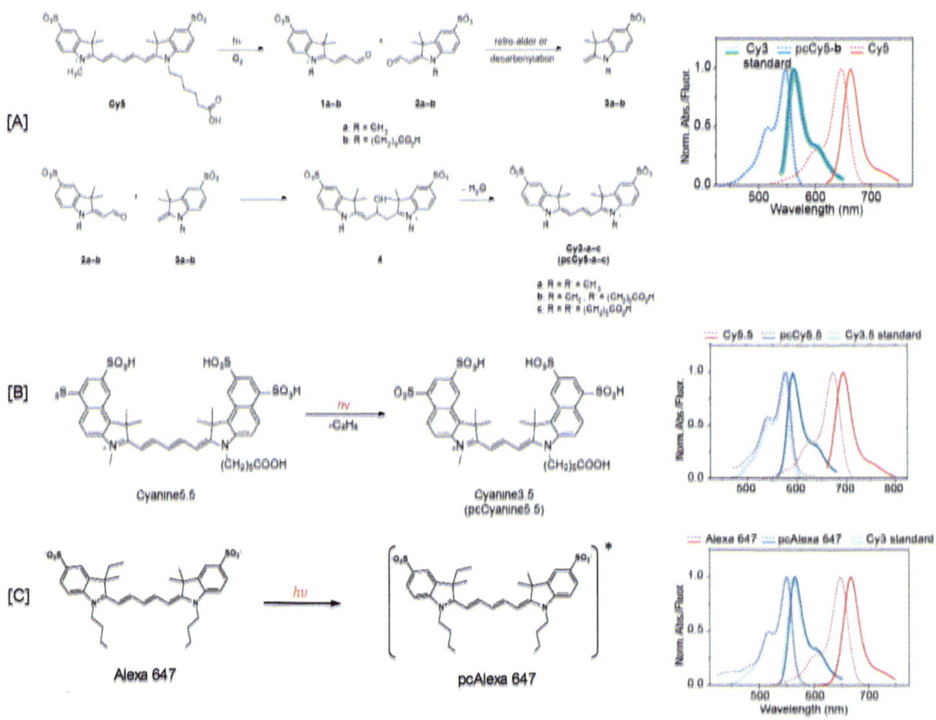

Fig. 7.8 (**a**) Chemical representation of Cyanine dyes, along with the photoisomerization mechanism showing conversion from neutral to .... form. (**b**) The excitation and emission spectra of Cy5, Cy5.5 and Alexa 647 are also shown. The figure is an adaptation from, Dmitriy M Chudakov et al., Tracking intracellular protein movements using photoswitchable fluorescent proteins PS-CFP2 and Dendra2, Nature Protoc. 2, 8 (2007)

supplementary notes. It is the loss of C_2H_2 unit that causes the absorption peak-shift in these dyes during photoconversion process. This shifts for Cy5, Cy5.5 and AF647 is shown in Fig. 7.8 (right panel). All other details of these far-red organic cyanine dyes are tabulated in Table 7.2.

Overall, it is quite clear that dyes have definite advantage over fluorescent proteins due to their brightness and extinction coefficient, among others. However, fluorescent proteins are preferred for prolonged live cell imaging which is largely due to their endogenous nature.

Table 7.2 Characteristics and switching properties of, Cy5, Cy5.5, Alexa Fluor 647. ∗ indicates the presence of GLOX and 10 mM MEA. The table is adapted from Ref. *Dempsey et al., Nature Meth. 8, 1027–1036 (20011)* and *Cho et al., J. Am. Chem. Soc. 143, 14125–14135 (2021)*

	Cy5 activation		Cy5.5 activation		AF647 activation	
	Before	After	Before	After	Before	After
Excitation	650 nm	547 nm	680 nm	580 nm	650 nm	547 nm
Emission	670 nm	–	694 nm	–	665 nm	–
Color	Red	Green	Red	Orange	Red	Green
Quantum Yield	0.28	–	0.28	–	0.20	–
Extinction Coeff. (M^{-1} cm^{-1})	250,000	–	250,000	–	239,000	–
Detected Photons per switching event	4254*	–	5831*	–	3823*	–

References

1. Varghese JM, Joshi P, Aravinth S, Mondal PP (2022) Single-Molecule Super-Resolution Microscopy Reveals Formation of NS2B3 Protein Clusters on Mitochondrial Network Leading to its Fragmentation during the Onset of Dengue (Denv-2) Viral Infection, BioRxiv. https://doi.org/10.1101/2022.12.18.520514
2. Varghese JM, Aravinth S, Pant N, Mondal PP (2023) The Anti-inflammatory Drug Leflunomide Inhibits NS2B3 Cluster Formation During Dengue Viral Infection as Revealed by Single Molecule Imaging, BioRxiv. https://doi.org/10.1101/2023.12.26.573168
3. Raut P et al (2022) Cetylpyridinium chloride (CPC) reduces zebrafish mortality from influenza infection: Super-resolution microscopy reveals CPC interference with multiple protein interactions with phosphatidylinositol 4,5-bisphosphate in immune function. Toxicol Appl Pharmacol 440:115913
4. Klein T, Proppert S, Sauer M (2014) Eight years of single-molecule localization microscopy. Histochem Cell Biol 141:561–575
5. Minoshima M, Kikuchi K (2017) Photostable and photoswitching fluorescent dyes for super-resolution imaging. J Biol Inorg Chem 22:639–652
6. Heilemann M, van de Linde S, Mukherjee A, Sauer M (2009) Super-resolution imaging with small organic fluorophores. Angew Chem Int Ed 48:6903–6908
7. Heilemann M, Margeat E, Kasper R, Sauer M, Tinnefeld P (2005) Carbocyanine dyes as efficient reversible single-molecule optical switch. J Am Chem Soc 127:3801–3806
8. Lin Y et al (2015) Quantifying and optimizing single- molecule switching nanoscopy at high speeds. PLoS ONE 10:e0128135
9. Lelek M et al (2021) Single molecule localization microscopy. Nat Rev Meth Prim 1:39
10. Hess ST, Girirajan TP, Mason MD (2006) Ultra-high resolution imaging by fluorescence photoactivation localization microscopy. Biophys J 91:4258–4272
11. Adam V et al (2008) Structural characterization of IrisFP, an opticalhighlighter undergoing multiple photo-inducedtransformations. PNAS 105:18343–18348

12. Fuchs J et al (2010) A photoactivatable marker protein for pulse-chase imaging with superresolution. Nat Methods 7:627–30
13. Colletier JP et al (2016) Serial femtosecond crystallography and ultrafast absorption spectroscopy of the photoswitchable fluorescent protein IrisFP. Jl Phys Chem Lett 7:882–887
14. Warren MM, Kaucikas M, Fitzpatrick A, Champion P, Sage JT, van Thor JJ (2013) Ground-state Proton Transfer in the Photoswitching Reactions of the Fluorescent Protein Dronpa. Nat Commun 4:1461
15. Grimm JB et al (2016) Bright photoactivatable fluorophores for single-molecule imaging. Nat Meth 13:985–988
16. Lippincott-Schwartz J, Patterson GH (2009) Photoactivatable fluorescent proteins for diffraction-limited and super-resolution imaging. Trends Cell Biol 19:555–565
17. Ando R, Hama H, Yamamoto-Hino M, Mizuno H, Miyawaki A (2002) Proc Natl Acad Sci USA 99:12651–12656
18. Chudakov DM, Lukyanov S, Lukyanov KA (2007) Biotechniques 42:553
19. McEvoy AL et al (2012) PLoS One 7:e51314
20. Dempsey WP et al (2015) Nat Methods 12:645–648
21. Mohr MA, Argast P, Pantazis P (2016) Nat Protoc 11:2419–2431
22. Uno SN et al (2014) A spontaneously blinking fluorophore based on intramolecular spirocyclization for live-cell super-resolution imaging. Nat Chem 6:681–689
23. Cubitt AB, Heim R, Adams SR, Boyd AE, Gross LA, Tsien RY (1995) Understanding, improving and using green fluorescent proteins. Trends Biochem Sci 20:448–455
24. Tsien RY (1998) The green fluorescent protein. Annu Rev Biochem 67:509–544
25. Palm GJ, Zdanov A, Gaitanaris GA, Stauber R, Pavlakis GN, Wlodawer A (1997) The structural basis for spectral variations in green fluorescent protein. Nat Struct Biol 4:361–365
26. Warren A, Zimmer M (2001) Computational analysis of Thr203 isomerization in green fluorescent protein. J Mol Graphics Modell 19:297–303
27. Ward WW, Bokman SH (1982) Reversible denaturation of Aequorea green-fluorescent protein: physical separation and characterization of the renatured protein. Biochemistry 21:4535–4540
28. Zimmer M (2002) Green Fluorescent Protein (GFP): applications, structure, and related photophysical behavior. Chem Rev 102:759
29. Chudakov DM et al (2007) Tracking intracellular protein movements using photoswitchable fluorescent proteins PS-CFP2 and Dendra2. Nature Protoc 2:8
30. Levitus M, Ranjit S (2011) Cyanine dyes in biophysical research: the photophysics of polymethine fluorescent dyes in biomolecular environments. Q Rev Biophys 44:12351
31. Cho Y et al (2021) Mechanism of cyanine5 to cyanine3 photoconversion and its application fot high-density single-particle tracking in a living cell. J Am Chem Soc 143:14125–14135

8 The Signature of a Single Molecule

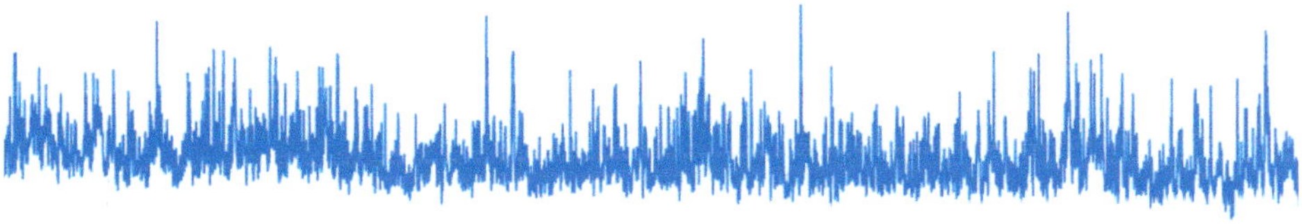

Distinguishing a single molecule from an ensemble requires a detailed understanding of a three-level system. The phenomenon underlying the emission from a single molecule depends on a number of factors, such as transition probabilities, emission rates, and selection rules associated with the energy levels in a three-level system. Among other factors, the quantum jump between two states (fluorescence and dark state) is a strong indicator of a single molecule. Such jumps cause a sudden burst of photons after a period of darkness at regular intervals (often equal to the lifetime of a molecular state such as, triplet state). Another key aspect elucidated by a single molecule is photon correlation, which occurs when the system jumps between singlet and triplet states, causing a strong correlation (during the ON cycle) or no photons for a brief period (OFF state). This is better illustrated for Pentacene molecule that exhibits all the phenomenons associated with a single molecule. As an example, Pentacene molecule is considered for case study. Infact, the last section is dedicated to the study of Pentacene molecule that elucidate key aspects of a single molecule.

The existence of single molecules and its experimental observation is debated for a very long time. However, this started to change in the late 1970s and 1980s, when Moerner's group gave the first conclusive evidence of emission from single molecule [1]. Although, some progress towards single molecule detection were made by Hirschfeld in the 1970s with the detection of labelled single antibody molecules, but did not get the attention due to lack of modern detectors (EMCCD, sCMOS, SPAD etc.). During the same time,

P. P. Mondal, S. Hess, *Single-Molecule Super-Resolution Optical Microscopy*,
https://doi.org/10.1007/978-3-032-13442-4_8

fluorescence correlation spectroscopy (FCS) also aided insight related to the dynamics of single molecules [2]. However, this changed in the early 1990s with the availability of low-light ultrasensitive detectors that has quantum efficiency approaching 1, low thermal noise, and high gain, among other factors.

Several attempts were made to detect single molecule during 1970s and 1980s. Finally, Moerner and Kador succeeded in the optical detection and spectroscopy of single chromophores in solids at low temperatures [1]. About the same time, Orrit and Bernard reported similar observations of fluorescence from single pentacene molecules [3]. The potential was quickly realized by other groups and the field saw a steady rise. Shera et al. reported the detection of single fluorophore under biologically relevant conditions [4]. Another significant advance was the translation from low to room temperature, achieved by Betzig and Chichester in the year 1993 [5]. This proved to be a big step towards studying biological systems. The mid-1990s saw the expansion of single molecule fluorescence both in biological and chemical systems. Some of these include, the direct observation of individual ATP turnovers during myosin movement [6], protein unfolding of single protein [7–9], direct visualization of its rotary motion provided by fluorescence imaging of F1-ATPase [10]. These developments have pushed forward the capability of modern microscopes and gave rise to a new kind of biology termed as, *Single Molecule Biology*. Some of the examples include, direct visualization of protein folding states, pathways of single protein and RNA, visualization of single enzyme, and popularly known hand-over-hand mechanism of myosin and kinesin.

Among others FCS has played a crucial role in the study of single molecules. FCS technique is used in all sub-domains of natural sciences [11, 12]. This is immediately relevent from the important developments in the subsequent years. Bresloff and Crothers carried out studies to measure fluorescence fluctuations caused by binding and unbinding of ethidium bromide (EB) to DNA [13]. In another experiment related to ED-DNA binding, the kinetics of chemical reactions was measured via fluctuations [2]. With the success of FCS on EB-DNA, the experiment was extended to the study of molecule mobility on cell surfaces on the then state-of-the-art confocal microscope [14]. Subsequently, W W Webb group used FCS to detect and track individual low density lipoprotein receptor (LDL-R) molecule and small receptor clusters with high precision of $\sim$30 nm [15]. In a series of experiments by Steve Block group in 1990s, it was shown that molecular motors (silica beads carrying single molecules of the motor protein kinesin on microtubules) move with regular steps of 8 nm using an optical tweezer interferometry [16, 17]. Therefter several experiments were performed using confocal microscopy to study single molecule (Rhodamine 700 and Cy5) and specifically the bursts of fluorescence photons [18, 19]. The year 1997 saw the impact of dual-color cross-correlation spectroscopy (FCCS) realized by Schwille et al. that showed direct measurement of molecular interactions with single molecule sensitivity [20, 21].

Then came the most exciting year 2006 that saw the development of single molecule localization microscopy. Three independent groups (Samuel Hess for University of Maine, Orono; Eric Betzig from Janelia Farm Research Campus in Ashburn, Virginia; Xiaowei

Zhuang, Harvard University, Massachusetts) demonstrated the power of single molecule imaging by creating a single molecule map of a biological specimen [22–24]. This gave rise to the most popular microscopy technique called *Single Molecule Localization Microscopy*. The invention of SMLM took natural science to the next level, much beyond the capabilities of existing diffraction-limited microscopes. Of all sciences, biology has captured much of attention and benefited from these amazing single molecule tools. To name a few, the last few years have seen protein dynamics in plasma membrane, signal transduction, clustering of non-structured viral proteins (Hemaglutinin) in plasma membrane (Influenza Type A) [25, 26], disruption of mitochondrial network promoted by viral protein accumulates (NS2B3) in type 2 Dengue [27], and protein-organelle interaction at varying stages of infection. All these cell based study of viral infection in disease biology (Influenza, and Dengue) has greatly enhanced our understanding of infectious diseases at single molecule level and may lead to potential drug treatment studies [28].

Although the foundations of single molecule investigations have been laid, the technology is in its infancy, and it is anticipated that this will bring revolution to natural sciences. Below we discuss some of the key experiments and the theory behind single molecule in a radiation field.

8.1 Fluorescence Correlation Spectroscopy for Single Molecules

Fluorescence autocorrelation spectroscopy is among the very first experiments that confirmed the presence of single molecules at room temperature. The technique is invented in the early 1970s by D. Magde, E. Elson, W.W. Webb [29–31]. In FCS, fluorescence is observed from fluorophores/labelled molecules (preferably in solution at very low concentrations, say picomolar) in a tiny detection PSF volume (typically few femtoliters), and correlation analysis is applied to recorded fluctuation in fluorescence intensity. Any physical process (such as, diffusion and photophysical transitions) that changes fluorescence intensity reflect as a decay in the correlation curve. Some of the examples include, molecule diffusing in and out of the detection volume/PSF, or molecule switching to non-fluorescent triplet state (intersystem crossing) giving triplet-to-ground state relaxation.

In fluorescence correlation spectroscopy, time-dependent fluorescence fluctuations are observed due to transit of single molecule in the detection volume/PSF. Poisson statistics at low concentrations predicts that the number of molecules in the femtoliter detection volume fluctuate predominantly between 0 and 1. This means, from the FCS data one can precisely say the time of entrance, stay duration and exit of single molecule in the volume. For example, Rhodamine 6G molecule in a solution form (at a concentration of 10 pM in water) gives fluorescence burst rates of upto 140 photons in 1 ms (about 140 kHz) as shown in Fig. 8.1a,b.

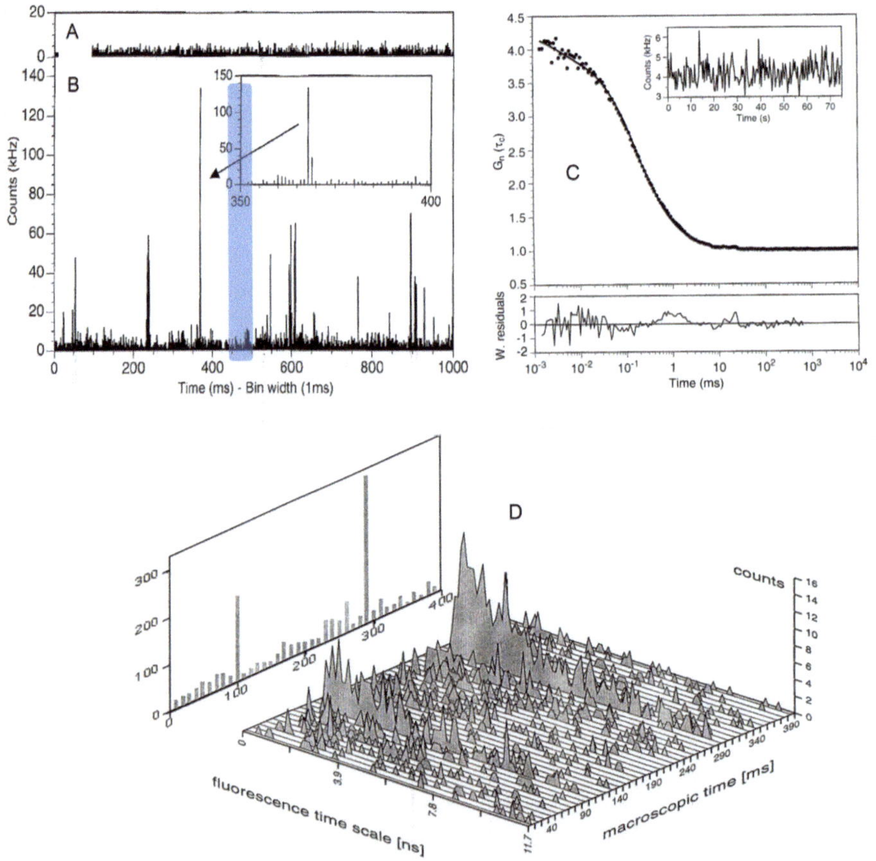

Fig. 8.1 Fluorescence from Rhodamine 6G molecules at a laser wavelength of 514 nm and intensity of 370 μW. Data acquisition with an integration time of 1 ms. (**a**) Rhodamine 6G (10 pM in water), (**b**) Pure water. (**c**) The normalized fluorescence autocorrelation function $G(\tau)$ with time trace (inset). (**d**) Fluorescence decay curves of bursts due to single-molecule transits. The inset shows time-integrated fluorescence signal trace (bin-width of 10 ms). The figure is adapted from, Zander et al., Applied Physics B 63, 517–523 (1996)

The three dimensional volume of the excitation laser intensity is assumed Gaussian,

$$I(r, z) = I_0 \, e^{-2r^2/w_0^2} \, e^{-2z^2/z_0^2} \tag{8.1}$$

where, I_0 is the amplitude. Here, $\frac{1}{e^2}w_0$ and $\frac{1}{e^2}z_0$ are the radius along radial (r) and axial (z) directions, respectively.

Assuming, that the fluorescence fluctuation is mainly caused by triplet decay and translational diffusion, the measured fluorescence $I(t)$ about average intensity, $\langle I \rangle$ can be expressed by auto-correlation function [32],

$$G(\tau_c) = \frac{1}{\langle I \rangle^2} \langle I(t+\tau_c)\, I(t)$$
$$= 1 + \alpha(1 - A + A\, e^{\tau_c/\tau_T}) + \left(1 + 4D\tau_c/w_0^2\right)^{-1} \left(1 + 4D\tau_c/z_0^2\right)^{-1/2} \quad (8.2)$$

where, $\alpha = \frac{(1-I_B/I)^2}{\sqrt{8}N(1-A)}$. Here, N is the average molecules in the PSF volume, A is the fraction of molecules in the triplet state with triplet correlation time τ_T, D is the diffusion coefficient, $\tau_C = \omega_0^2/4D$ is the characteristic time for diffusion, and I_B is the background. Figure 8.1c shows a typical fluorescence correlation function of 10^{-11} M Rhodamine 6G solution. The bursts of single molecules with fluorescence time scale (ns) and macroscopic time scale is also shown in Fig. 8.1d.

A fit to the curve with Eq. 8.2 (using Marquardt-Levenberg least squares algorithm) revels an average of $N = 0.033$ molecules. This indicates that FCS detects predominantly single-molecule events. So, fluorescence autocorrelation spectroscopy allows the detection of single molecules at room temperature. Overall, FCS along with many other independent experiments suggests that the excitation and detection of single molecule strongly depends on light-molecule interaction. We will explore this in the rest of the chapter.

8.2 Interaction Between Light and Single Molecules (Theory)

In general, interaction of a single molecule with a field involves field dipole interaction [33]. Accordingly, the interaction Hamiltonian (H_i) between the molecule and the incident field is given by,

$$H_i = \vec{\mu}.\vec{E} \quad (8.3)$$

where, $\vec{\mu} = q.\vec{r}$ is the transition dipole moment, with q as the charge, and r as the distance from the centre.

Now, consider a three-level system (representing energy levels in a single molecule) as shown in Fig. 8.2. In general, the photophysical dynamics of a single molecule under coherent excitation has two major components. For a coherent motion driven by light field, there exists Rabi oscillations between singlet ground and excited states (represented

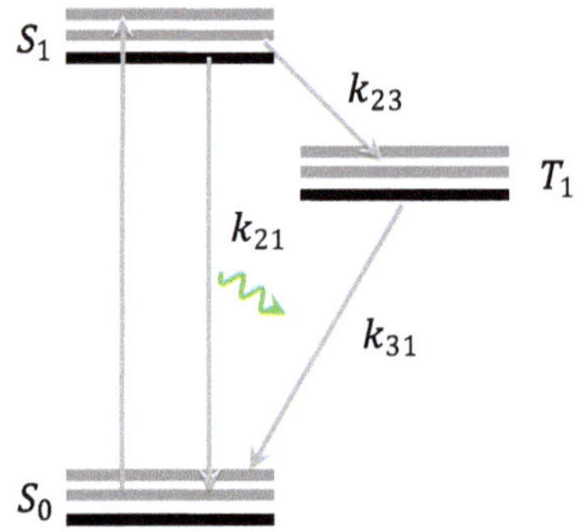

Fig. 8.2 Energy level diagram for a typical three-level fluorescent molecule. The energy levels are ground singlet state (S_0), first excited singlet state (S_1), and metastable triplet state T_1. The corresponding rate equations are also mentioned

by density matrix, σ). Second, the incoherent transitions are responsible for population transfer between electronic levels. The corresponding density matrix is given by,

$$\begin{bmatrix} \sigma_{11} & \sigma_{12} & \sigma_{13} \\ \sigma_{21} & \sigma_{22} & \sigma_{23} \\ \sigma_{31} & \sigma_{32} & \sigma_{33} \end{bmatrix} \tag{8.4}$$

where, the diagonal elements are real numbers that sum to one (also called populations of the states, $|1\rangle$ (or S_0), $|2\rangle$ (or S_1), $|3\rangle$ (or T_1)). The off-diagonal elements are complex conjugates of each other (also called coherences). In short, coherences in a mixed state of the density matrix (represented by off-diagonal elements) provide information about interference between two states [34].

Note that, a density matrix (or density operator) is a matrix whose elements (diagonal and off-diagonal) are used to calculate the probabilities of the outcomes of a measurement performed on a physical quantum system. The states represented by diagonal elements are pure states while those represented by off-diagonal elements are mixed states. Loosely speaking, mixed state is a kind of superposition state. In these systems, mainly two situations arise: (1) randomly produced pure states, and (2) a state entangled with another state. This happens for a state interacting with some environment (e.g., decoherence). The combined result would be statistically same upon measurement.

Following the work of deVries and Wiersma [34], the equation of motion for the density matrix elements of a three-level system (under electric dipole and rotating-wave approximation) are given by Bernard et al. [35],

$$\begin{cases} \dot{\sigma_{11}} = k_{21}\sigma_{22} + k_{31}\sigma_{33} + \frac{i\Omega}{2}(\sigma_{21} - \sigma_{12}) \\ \dot{\sigma_{22}} = -(k_{21} + k_{23})\sigma_{22} + \frac{i\Omega}{2}(\sigma_{12} - \sigma_{21}) \\ \dot{\sigma_{33}} = k_{23}\sigma_{22} - k_{31}\sigma_{33} \\ \dot{\sigma_{12}} = \frac{i\Omega}{2}(-\sigma_{11} + \sigma_{22}) + (i\Delta - \Gamma_2)\sigma_{12} \end{cases} \tag{8.5}$$

with the condition, $\sigma_{11} + \sigma_{22} + \sigma_{33} = 1$. Here, $\Omega = \frac{1}{\hbar}|\vec{\mu}_{12}.\vec{E}|$ is the Rabi-frequency[1] associated with spontaneous transitions, $\hat{\mu}_{12}$ is the transition dipole moment between states, $|1\rangle$ and $|2\rangle$, and $\vec{E}$ is the light field. Γ_2 is the dephasing rate[2] of $S_0 - S_1$ coherence, k_{ij} are the transition rates, and Δ is the detuning factor i.e, detuning of the laser with respect to molecular resonance frequency.

Here, we are interested in steady-state solution that enables the determination of line-shape and fluorescence emission rate. So, the solution to the set of Eqs. 8.5 is obtained by setting the following condition,

$$\begin{bmatrix} \dot{\sigma_{11}} \\ \dot{\sigma_{22}} \\ \dot{\sigma_{33}} \\ \dot{\sigma_{12}} \end{bmatrix} = \begin{bmatrix} k_{21}\sigma_{22} + k_{31}\sigma_{33} + \frac{i\Omega}{2}(\sigma_{21} - \sigma_{12}) \\ -(k_{21} + k_{23})\sigma_{22} + \frac{i\Omega}{2}(\sigma_{12} - \sigma_{21}) \\ k_{23}\sigma_{22} - k_{31}\sigma_{33} \\ \frac{i\Omega}{2}(-\sigma_{11} + \sigma_{22}) + (i\Delta - \Gamma_2)\sigma_{12} \end{bmatrix} = 0 \tag{8.6}$$

The above set of equations can be solved using Laplace transformation and applying the condition, $\sigma_{11} = 1$, $\sigma_{22} = \sigma_{33} = \sigma_{12} = 0$ at time $t = 0$ as discussed in Ref. [34]. For steady state, the expression for population difference (from ground to excited electronic state, $|1\rangle \rightarrow |2\rangle$) is given by,

$$\sigma_{11} - \sigma_{22} = \frac{\Delta^2 + 1/\tau_2^2}{\Delta^2 + 1/\tau_2^2 + \kappa^2} \tag{8.7}$$

where, $\kappa^2 = \chi^2(2 + r)/2\tau_2 k_T$, with, $k_T = k_{21} + k_{23}$ as the total decay constant of level $|2\rangle$, $r = k_{23}/k_{31}$ as the ratio of transition rates, and $\Gamma_2 = 1/\tau_2$ is the dephasing rate of $S_0 - S_1$ coherence.

In addition, Eq. 8.6 gives the expression for excitation line-width ($\Delta\nu$) and the emission-rate as [36],

$$\begin{cases} \Delta\nu(I) = \Delta\nu(0)\sqrt{1 + I/I_s} \\ R(I) = \frac{1}{\tau_f}\,\phi_F\,\sigma_{22} = R_\infty \frac{I/I_s}{1 + I/I_s} \end{cases} \tag{8.8}$$

[1] When exposed to electromagnetic field (such as light), atoms does not just jump-up to the excited state, instead it begins to oscillate between the two states. The cycle repeats indefinitely as long as it is exposed to coherent light field. The frequency at which the oscillation occurs is called Rabi frequency.

[2] A molecule excited by a short pulse places it in coherent super-position state, $|\psi(t)> = C_0|S_0> +C_1|S_1>$. The $S_0 - S_1$ coherence is represented by off-diagonal density matrix elements. Due to slightly different chemical environment around single molecules, the $S_0 - S_1$ energy-gap fluctuate in time, causing slightly different phase drift from ensemble coherence decay. This contribute to dephasing.

where, $\tau_f = 1/k_T$ is the fluorescence lifetime, and ϕ_F is the fluorescence quantum yield. Here, I and I_s are respectively the intensity and saturated intensity.

One can also derive the expression for saturation intensity and emission rate. Following Ambrose et al. [36], using the density matrix-treatment, the expression for saturation intensity is given by,

$$I_s = \frac{\epsilon_0 c\hbar^2 (k_{21} + k_{23})}{|\vec{\mu}_{21}|^2 (2 + A)\tau_2} \tag{8.9}$$

and the saturated fluorescence emission rate is,

$$R_\infty = \frac{(k_{21} + k_{23})\,\phi_F}{2 + A} \tag{8.10}$$

where, $A = k_{23}/k_{31}$.

The experimental data fitted by the expression for excitation line-width and the fluorescence emission rate (Eq. 8.8) from a single Pentacene molecule is shown in Fig. 8.3. This gives an approximate estimate for line-width of 7.3 ± 0.8 M Hz, and a fully saturated photon count-rate of, 8100 ± 300 counts/sec giving, $R_\infty = (7.2 \pm 0.2) \times 10^5$ photons/sec. Alternately, the values of these parameters can be verified from photophysical experiments on an ensemble of molecules [37–41]. Noting that, the rates $k_{21} + k_{23} = 4.2 \times 10^7\ s^{-1}$,

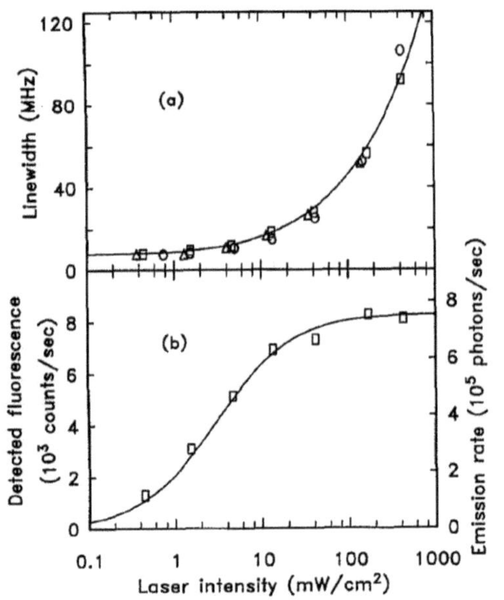

Fig. 8.3 Saturation behavior of a single pentacene molecule. (**a**), (**b**) The fluorescence excitation line-width and fluorescence emission rate as a function of laser intensity at a temperature of $1.5K$

$k_{23} \sim 10^3\ s^{-1}$, $k_{31} \sim 10^3\ s^{-1}$ [42], and a fluorescence yield of 0.08, the calculated saturation emission rate, $R_\infty = 11.2 \times 10^5$ photons s^{-1}. In addition, detector efficiency, sensitivity, and optical losses (reflection, absorption) reduce this by a factor of 100, leading to photon detection in the range of ~ 1000.

These experimental values strongly indicate that the emission from a single molecule is significant, but needs sophisticated detection device to detect, given the photon background in laboratory conditions ($\sim 10^2$ photons). In order to overcome the background and other potential sources that aid to the background, low temperatures and dark conditions are preferred. Low temperature directly minimize thermal and read noise, apart from reducing other sources of electronic noises.

Noting that single molecules can't be ON for indefinite time, so there is a characteristics time-scale for emission and the period of darkness. This will be the subject matter for the next subsection. In addition, we will discuss the mechanism behind single molecule blinking, and the number of photons emitted during the ON period.

8.3 Quantum Jumps in a Trapped Ion/Single Molecule

One of the key signatures of single molecule is its ability to blink, and emit a burst of photons followed by a prolonged period of darkness (no photons). However, the number of emitted photons depends on the molecule, and its immediate chemical neighborhood. The key lies on the state of molecule, the transiton rates between energy levels, selection rules and fundamentally on the quantum state of energy levels. Understanding single molecule blinking will be the subject matter of this section.

Some of the early experiments goes back to the year 1989 from the work of W. E. Moerner demonstrating the optical detection and spectroscopy of single molecules in condensed phase [43]. In those days, the experiments are performed on simple quantum systems such as, atoms or ions in vapor or single molecules trapped in a crystal at ultra-low temperatures or single molecules in optical traps (such as, Paul trap). Since these quantum systems are fairly well understood theoretically, it was felt likely to give experimentally convincing results. Actually, the concept of quantum jump dates back to 1975, when Dehmelt proposed that the quantum state of a molecule/single ion (in a V-configuration) could be monitored from a fluorescent molecule [44, 45]. In this scheme, a molecule is resonantly driven in the allowed transition where strong fluorescence is observed. Eventually, the electron will undergo a transition to metastable state, although with small probability. Since the molecule in metastable state live longer (on the order of its lifetime), there will be no photon emission during this period. At this stage, a single quantum jump to the ground state will put back the molecule in the resonant fluorescence cycle (allowed transition), triggering a strong photon emission. Thus a repeat of these transitions will give rise to a sequence of bright and dark intervals, commonly known as *blinking*. Later on, blinking is experimentally observed by three different groups that ultimately became the basis of single molecule imaging and microscopy [46–48]. However, note that there exists several mechnaism that may give rise to blinking. Henceforth, we will focus our attention to single molecule blinking, its dynamics and associated properties.

Similar to atomic systems, molecules have discrete molecular energy levels labelled as singlet (such as, S_0, S_1, ...) and triplet (T_1, and $T_{n\geq 2}$). According to Fig. 8.2 that comprises of ground and excited energy levels, the time distribution of photons emitted by a single molecule show bright intervals is separated by dark periods. In the emission cycle ($S_0 \longleftrightarrow S_1$) involving the ground singlet S_0 and excited singlet S_1 energy levels, photons continuously emit at near-regular intervals before being shelved in the metastable triplet state (T_1). This cause the photons to be bunched in time, also called *photon bunching*. After a dark period (of the order of triplet state lifetime), a weak transition from T_1 to S_0 will trigger an avalanche of fluorescence photons due to the activation of fluorescence cycle ($S_0 \longleftrightarrow S_1$). It would be great to observe these intermittent ON/OFF of single molecules. This requires a number of factors including sensitive detection, detection rate, background/noise reduction, and photophysical properties of the single molecule.

To understand, we take the example of a single terrylene molecule in p-terphenyl with known transition rates, $k_{23}{\sim}10^3\,\mathrm{s}^{-1}$, and $k_{31}{\sim}10^3\,\mathrm{s}^{-1}$ [49]. The transition rates indicates that a single terrylene molecule stays in the single system or fluorescence cycle ($S_0 \longleftrightarrow S_1$) for about $2/k_{23} \approx 2\,\mathrm{ms}$ and after intersystem crossing into the triplet state it will be shelved for $1/k_{31} \approx 1\,\mathrm{ms}$. Considering a typical fluorescence photon of $\approx 500{,}000\,\mathrm{s}^{-1}$ of a single terrylene molecule at saturation intensities, fluctuations in the fluorescence can be observed, given a high time sampling (say, $100\,\mu\mathrm{sec}$) and a high quantum efficiency of the detector (say, >90%). For example, an average fluorescence from a single terrylene molecule is shown in Fig. 8.4. It is evident that the single molecule jumps between singlet and triplet states (which is an intramolecular process). So, the distribution of interval lengths for ON times (during singlet-singlet cycle), and OFF time (during its stay in triplet state) would allow us to determine the states of the underlying process. A typical such distribution is shown in Fig. 8.5. To determine the rates of ON and OFF time-intervals, exponentials are fitted. It is clear from the data that a single exponential fits well for ON data, whereas OFF data requires bi-exponential (one at short times, here 0–3 ms, and the other at relatively larger times i.e, $> 3\,\mathrm{ms}$). From the fits to the distribution data, $\tau_{ON} = 6.3\,\mathrm{ms}$, and $\tau^1_{OFF} = 0.43\,\mathrm{ms}$, $\tau^2_{OFF} = 3\,\mathrm{ms}$. Here, $\tau_{ON} = 2/k_{23}$ is the lifetimes of the sub-level $2 \rightarrow 3$ (intersystem crossing) when $S_0 \rightarrow S_1$ transition is saturated (see, Fig. 8.5). The other lifetimes, $\tau^1_{OFF} = 0.43\,\mathrm{ms}$ and $\tau^2_{OFF} = 3\,\mathrm{ms}$ are essentially rates for $3 \rightarrow 1$ transitions. The rates are determined by a separate experiment involving the measurement of fluorescence autocorrelation function. The rate values are found to be, $k_{23} = 0.4 \times 10^3\,\mathrm{s}^{-1}$, $k^{(1)}_{31} = 2.3 \times 10^3\,\mathrm{s}^{-1}$, and $k^{(2)}_{31} = 0.3 \times 10^3\,\mathrm{s}^{-1}$, respectively. The stochastic discrete jumps between different electronic levels is quite evident and so a single terrylene molecule qualifies as a single quantum system evolving with time. Here, the jumps are between single and triplet states which is directly observed as a discrete

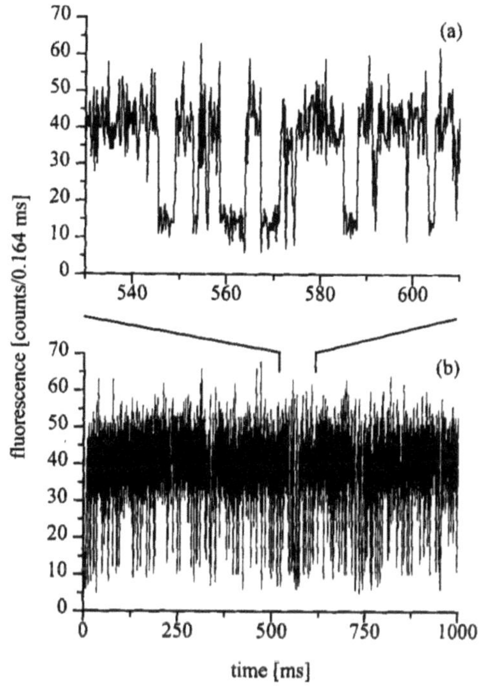

Fig. 8.4 Fluorescence counts detected for a single terrylene molecule (in p-terephenyl) as a function of time. (**b**) A small section (1 sec) of a larger data set (26.8 secs), (**a**) A 80 ms data is displayed on a larger scale. The figure is adapted from, T. Basche et al., Book: Single molecule optical detection, imaging, and spectroscopy, VCH pub., 1996

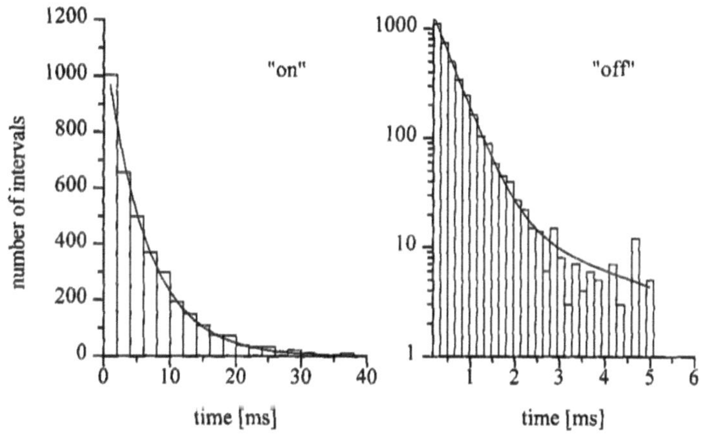

Fig. 8.5 The distribution of ON and OFF intervals of a single terrylene molecule. The source data is a trace of 26.8 sec data, of which a small section (see, Fig. 8.4) is shown. The figure is adapted from, T. Basche et al., Book: Single molecule optical detection, imaging, and spectroscopy, VCH pub., 1996

interruptions of strong fluorescence signal. Thus, in-principle, emission from a single molecule is a true quantum mechanical system which were traditionally restricted to a single atom.

It is evident that, quantum jump is an essential characteristics of a single molecule. Specifically, this translates to the blinking of a single quantum system involving single-triplet quantum jumps.

8.4 Photon Correlations from a Single Molecule

A consequence of quantum jumps in single molecules lead to photon correlations. The jumps between singlet and triplet states suggests an emission pattern from single molecules where the photon emissions are either strong or completely missing for a brief period of time. Experimentally, photon correlations are better manifested in fluorescent molecules undergoing singlet-triplet transitions. A simple fluorescence counting experiment would enable us to study photon correlations in time. Indirectly, the time distribution of emitted photons and their correlation is the reflection of allowed photophysical transitions and the associated intermolecular dynamics. By recording the fluorescence intensity auto-correlation function, it is possible to determine dynamical processes from nanosecond to millisecond time scale that are strongly associated with the lifetime of various excited states (singlet or triplet) of a single molecule.

Thus, it becomes imperative to study emission from a single molecule and the temporal photon correlation. To no ones surprise, we will use the correlation theory for photon emission from single molecules. The theory is taken from the work of Loudon who used it for understanding correlation for classical and non-classical quantum light fields [50]. Accordingly, the correlation function of the field, $E(t)$ (first order correlation function, $g^{(1)}(\tau)$) and of the intensity (second order correlation function, $g^{(2)}(\tau)$) are given by,

$$\begin{cases} g^{(1)}(\tau) = \frac{1}{\langle E^2(t)\rangle} \langle E(t)\, E(t+\tau)\rangle \\ \\ g^{(2)}(\tau) = \frac{1}{\langle I(t)\rangle^2} \langle I(t)\, I(t+\tau)\rangle \end{cases} \tag{8.11}$$

where, the explicit expression for numerators are,

$$\begin{cases} \langle E(t)\, E(t+\tau)\rangle = \lim_{T\to\infty} \frac{1}{T} \int_0^T E(t)\, E(t+\tau)\, dt \\ \langle I(t)\, I(t+\tau)\rangle = \lim_{T\to\infty} \frac{1}{T} \int_0^T I(t)\, I(t+\tau)\, dt \end{cases} \tag{8.12}$$

Experimentally, photon correlation is measured from the detection of emitted photons by a sensitive detector (e.g., photo-multiplier tube) and the arrival time is recorded for each event (refers to photon detection). In the fluorescence mode, photon count is the fluorescence intensity at time t i.e, $I(t)$, and counting of photon-pairs separated in time by τ is the fluorescence intensity auto-correlation function, $g^{(2)}(\tau)$. Here, we will follow the work of Orrit to determine the correlation function for single molecule [35, 51]. Accordingly, the probability of detecting a photon-pair at time t is the probability of occupation of excited singlet state at t i.e, $P(S_1, t)$ and the conditional probability of molecule in the same state at time $t + \tau$ i.e, $P(S_1, t + \tau | S_0, t)$, since the emission of the first photon and its subsequent observation projects the system into its ground state, S_0. The conditional probability is proportional to the population of the singlet excited state i.e, $P(S_1, t + \tau | S_0, t) \propto \sigma_{22}(\tau)$. Here, $\sigma_{22}(\tau)$ is the solution of Block equations for a 3-level system (Eq. 8.5). Following, Ref. [51], the intensity correlation can be shown to be,

$$g^{(2)}(\tau) = \frac{1}{P(S_1, t)^2}[P(S_1, t)\ P(S_1, t + \tau | S_0, t)] = \frac{\sigma_{22}(\tau)}{\sigma_{22}(\infty)} \tag{8.13}$$

where,

$$P(S_1, t + \tau | S_0, t) = \eta\, k_{21}\, \sigma_{22}(\tau) \tag{8.14}$$

with η the detector efficiency, k_{21} is the radiative $S_1 \rightarrow S_0$ transition rate, and $\sigma_{22}(\infty)$ is the equilibrium population of the excited singlet state S_1.

Following the work of Bernard and Orrit, the explicit expression for $\sigma_{22}(\tau)$ is obtained by solving the set of equations (Eq. 8.5) using Laplace transform [35]. Accordingly, the the population for the first singlet excited state S_1 can be expressed as,

$$\sigma_{22}(\tau) = \frac{\Omega^2}{2}\left[\frac{k_{31}\Gamma_2}{\prod_i \lambda_i} + \sum_{i=1}^{4} \frac{(\lambda_i + k_{31})(\lambda_i + \Gamma_2)}{\lambda_i(\lambda_i - \lambda_j)(\lambda_i - \lambda_k)(\lambda_i - \lambda_l)} e^{-\lambda_i t}\right] \tag{8.15}$$

where, Ω is the Rabi frequency, Γ_2 is the rate of dephasing of $S_0 - S_1$ coherence and λ_i are the solutions of a 4th degree polinomial representing decay rates [35,51].

The Eqs. (8.13–8.15) gives the explicit expression for correlation function that can be used to determine photons correlations from a single molecule at varying time-scales (short, mid-range, and long). This is the subject matter of the next section for a known molecule, *Pentacene*.

8.4.1 Properties of Correlation Functions

The correlation function (also known as the **fluorescence intensity autocorrelation function**), $g^{(2)}(\tau)$ is an important indicator of light bunching and anti-bunching. In short, this function determines if the photons in a beam stream are correlated, anti-correlated or uncorrelated.

Explicitly, the function $g^{(2)}(r, t;\ r, t + \tau)$ is defined as the correlation between between photoevents at a location r (in the detector chip), but at different time-points, t and $t + \tau$. Experimentally, measurements at two different time-points is achieved by splitting the light into two parts using a beam-splitter, followed by delaying one part by τ and recombining both the parts on a sensitive detector (with high QE). Accordingly, the following three distinct cases arises for $r_1 = r_2 = r$ (say):

$$\begin{cases} g^{(2)}(\tau) = 1, \ uncorrelated \\ g^{(2)}(\tau) > 1, \ correlated \\ g^{(2)}(\tau) < 1, \ anti-correlated \end{cases} \tag{8.16}$$

Each of these cases are unique and represent distinct distribution of photons in time. For all the cases, we have assumed that the photons are arriving at a single location (r), and all the photons emitted by a single molecule are detected. For $g^{(2)}(r, t;\ r, t+\tau) > 1$, there is a non-zero positive chance of detecting a photon at t and a second photon at $t+\tau$, i.e, events are positively correlated. So, the likelihood of occurrence of one event makes the other event more likely, suggesting bunching of photons in time. However, $g^{(2)}(r, t;\ r, t+\tau) < 1$ ensures less likely of the occurrence of event (detection of a second photons at a time $t+\tau$) if an event (detection of first photon) occurred at t. This suggests photons distance from each-other in time and prefer to be detected one at a time. Finally, for $g^{(2)}(r, t;\ r, t+\tau) = 1$ indicates that emissions in time are unbunched since emissions occur independently and in complete randomness. A detailed discussion on this topic can be found in [52]. Next we employ correlation theory to understand photon emission from a single Pentacene molecule.

8.4.2 Case Study of a Pentacene Molecule

The derivation of exact expression for $\sigma_{22}(t)$ (Eq. 8.15) and the corresponding fluorescence intensity auto-correlation (Eq. 8.13) enables us to computationally study few salient features of the photo-physics of a known molecular system (Pentacene molecule) [53].

To determine $g^{(2)}(\tau)$ function, it is assumed that the Pentacene molecule is trapped in a solid matrix at low temperatures. In addition, it is a 3-level system and the photophysical parameters are known (see, previous section). The computational simulation of the correlation function ($g^{(2)}(\tau)$) for Pentacene molecule from nano-second to milli-second time is carried out. Figure 8.6 shows the characteristics of $g^{(2)}(\tau)$ function for varying values of Rabi-frequency, Ω. Broadly, the plot can be segregated in three timescales : Short times ($< 10^{-8}$ sec), Mid-range ($10^{-8} - 10^{-5}$ sec) and Long times ($> 10^{-5}$ sec). These range signify different characteristics of the photon correlation associated with a single molecule.

Short times ($< 10^{-8}$ sec) In this range, $g^{(2)}(\tau) \rightarrow 0$ i.e, the photon correlation is nearly zero. This indicates that the emission of second photon is not possible just after the emission of first photon. So the probability of observing photon-pair is zero at $\tau \rightarrow 0$. This also suggests non-classical light or quantum light, a characteristics of quantum field called photon-antibunching (tendency of photons to space-out in time). The reason behind

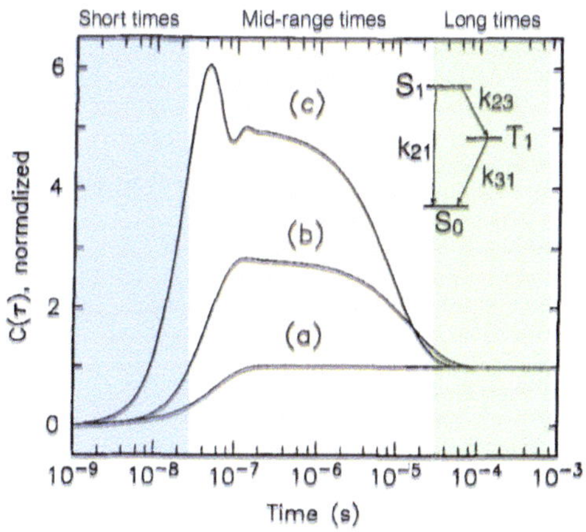

Fig. 8.6 Computational study of the correlation function, $g^{(2)}(\tau)$ for a single pentacene molecule at varying Rabi frequencies, (**a**) 3 MHz, (**b**) 25.5 MHz, (**c**) 71.3 MHz. Experimentally measured photo-physical parameters are used for the study. The figure is taken from, Phys. Rev. Lett. 69, 1516, (1992)

this nature lies in the fact that upon emission of a photon, a single molecule is in ground state and emission cannot happen from ground state. This requires a finite time (half of Rabi period) to elapse before molecule can again be in the excited state. This behavior (photon antibunching) is observed in several experiments [54–56].

Rabi-Oscillations (Peak) The general expression for $g^{(2)}(\tau)$ can be determined for specific case of short times when the contribution from triplet state is negligible. For a typical single molecule, the expression is given by Basche et al. [53],

$$g^{(2)}(\tau) = 1 - \left[\left(\frac{k_{21} + \Gamma_2}{2\Omega}\, sin(\Omega\tau) + \, cos(\Omega\tau)\right)\right] e^{-(k_{21}+\Gamma_2)\tau/2} \tag{8.17}$$

This clearly shows the sinusoidal nature of trigmonetric functions (sin and cos) that is responsible for oscillations also known as *Rabi Oscillations*. This happens at short times, represented by sharp peak. Note the dephasing rates, $\Gamma_2 = 1/T_2$ and $1/T_1$ as the damping terms. In the case of large intensity, the correlation function displays Rabi oscillations which is a laser driven coherent motion between singlet ground and excited state of a fluorescent single molecule. These oscillations are damped by phase relaxation processes.

Note that, Rabi oscillations exist in a two-level system (system with two transition states, ground and excited) when it is driven by an external field. When the light interacts with the molecule (assuming two states (S_0 and S_1) in the fluorescent cycle), the energy is exchanged between them. This energy exchange is periodic in nature and forms oscillations called *Rabi oscillations*. So, molecules in fluorescence cycle exhibit cyclic behavior when in resonance with the oscillatory driving field (also called Rabi cycle).

Mid-range times ($10^{-8} - 10^{-5}$ sec) This is the region where the molecule jumps between singlet and triplet states giving rise to photon bunching. This appears as monotonic decay (post peak in Fig. 8.6) of the $g^{(2)}$ function in the mid-range. In this relatively long time range, all the coherences have vanished and thus the correlation function (Eq. 8.15) can be shown to assume the following expression [57],

$$g^{(2)}(\tau) = 1 + C_1\, e^{-\lambda t} \tag{8.18}$$

where, the decay rate λ has contributions from k_{23} and k_{31} (rate of triplet state) given by,

$$\lambda = k_{31} + k_{31}\frac{I/I_s}{(1 + 2k_{31}/k_{23})}\left[1 + \frac{2k_{31}(I/I_s)}{(k_{23} + 2k_{31})}\right] \tag{8.19}$$

and the contrast C_1 is given by,

$$C = \left(\frac{\lambda}{k_{31}} - 1\right) \tag{8.20}$$

In this regime, the decay of the correlation function is caused by singlet-triplet transitions. During the fluorescence cycle ($S_0 \leftrightarrow S_1$), the molecule continue to emit photons at a rate given by fluorescence lifetime. However, no photons are emitted when the molecule makes a transition to triplet state. Due to the large lifetime of the triplet state, the molecule stays in the state without emitting photon. This results in the bunching of photons where bursts of emissions (observed during the fluorescence cycle) are separated by dark intervals.

Long times ($> 10^{-5}$ sec) At very long times, $g^2(\tau)$ reaches the minimum value of 1 that correspond to uncorrelated events. This is equivalent to the state when the molecule exhibits Poissionian emission. In short, Poissionian emission is a process where the emission or non-emission of photon does not effect the subsequent emissions from the molecule. From Fig. 8.6, it is clear that the contrast in the correlation function due to photon bunching vanishes at low excitation intensities. However, the average time spacing between fluorescence photons approaches the length of dark period (which equals the lifetime of metastable triplet state) at relatively low intensity.

It is quite evident that the same single molecule exhibits both antibunching (for small times) and bunching (at relatively large times). However, at very long times, the photons emission events are random and uncorrelated.

8.5 Summary of Single Molecule Properties

Based on the above description, it is quite apparent that single molecule have a distinct properties as compared to the bulk. Here, we sum up some of the key properties of single molecule:

1. Quantum Jumps: The fact that, single molecules jumps between fluorescent singlet state ($S_0 \leftrightarrow S_1$) and dark triplet states giving intermittency, isolates it from bulk and gives it a distinct characteristics.

2. Photon Bursts: On excitation (within the fluorescence cycle), single molecules emit a large burst of photons before entering the dark state. The number of photons in these bursts are an essential characteristics of a single molecule and can be use to distinctly identify/fingerprint them (in a similar environment). However, the photon emission characteristics change when exposed to different chemical environment.
3. Photon correlation: The photon emission from a single molecule may be bunched, antibunched or unbunched. The timescale of bunching/antibunching within a blinking period is specific to each molecule.

With the knowledge of single molecule, its emission and related characteristics, we are in a position to explore it in natural sciences. Specifically, in microscopy, these optical characteristics are better realizable and implementable, allowing visualization of single biomolecules in living systems (cells, organisms etc).

References

1. Moerner WE, Kador L (1989) Finding a single molecule in a haystack—optical-detection and spectroscopy of single absorbers in solids. Anal Chem 61:A1217–A1223
2. Magde D, Elson EL, Webb WW (1972) Thermodynamic fluctuations in a reacting system - measurement by fluorescence correlation spectroscopy. Phys Rev Lett 29:705–708
3. Orrit M, Bernard J (1990) Single pentacene molecules detected by fluorescence excitation in a para-terphenyl crystal. Phys Rev Lett 65:2716–2719
4. Shera EB et al (1990) Detection of single fluorescent molecules. Chem Phys Lett 174:553–557
5. Betzig E, Chichester RJ (1993) Single molecules observed by near-field scanning optical microscopy. Science 262:1422–1425
6. Funatsu T, Harada Y, Tokunaga M, Saito K, Yanagida T (1995) Imaging of single fluorescent molecules and individual ATP turnovers by single myosin molecules in aqueous-solution. Nature 374:555–559
7. Kellermayer MSZ, Smith SB, Granzier HL, Bustamante C (1997) Folding—unfolding transitions in single titin molecules characterized with laser tweezers. Science 276:1112–1116
8. Rief M, Gautel M, Oesterhelt F, Fernandez JM, Gaub HE (1997) Reversible unfolding of individual titin immunoglobulin domains by AFM. Science 276:1109–1112
9. Tskhovrebova L, Trinick J, Sleep JA, Simmons RM (1997) Elasticity and unfolding of single molecules of the giant muscle protein titin. Nature 387:308–312
10. Noji H, Yasuda R, Yoshida M, Kinosita K (1997) Direct observation of the rotation of F-1-ATPase. Nature 386:299–302
11. Thompson NL (2002) Fluorescence correlation spectroscopy Topics in Fluorescence Spectroscopy. Springer, Berlin, pp 337–378
12. Krichevsky O, Bonnet G (2002) Fluorescence correlation spectroscopy: the technique and its applications. Rep Prog Phys 65:251
13. Bresloff JL, Crothers DM (1975) DNA-ethidium reaction kinetics: demonstration of direct ligand transfer between DNA binding sites. J Mol Biol 95:103–123
14. Koppel DE, Axelrod D, Webb WW et al (1976) Dynamics of fluorescence marker concentration as a probe of mobility. Biophys J 16:1315–1329

15. Ghosh RN, Webb WW (1994) Automated detection and tracking of individual and clustered cell surface low density lipoprotein receptor molecules. Biophys J 66(5):1301–18
16. Svoboda K, Schmidt CF, Schnapp BJ, Block SM (1993) Direct observation of kinesin stepping by optical trapping interferometry. Nature 365:721–7
17. Schnitzer MJ, Block SM (1997) Kinesin hydrolyses one ATP per 8-nm step. Nature 388:386–90
18. Sauer M, Drexhage KH, Zander C, Wolfrum J (1996) Diode laser based detection of single molecules in solutions. Chem Phys Lett 254(3–4):223–228
19. Zander C, Sauer M, . . . et al (1996) Detection and characterization of single molecules in aqueous solution. Appl Phys B 63:517–523
20. Schwille P, Meyer-Almes FJ, Rigler R (1997) Dual-color fluorescence cross-correlation spectroscopy for multicomponent diffusional analysis in solution. Biophys J 72:1878–1886
21. Schwille P, Haupts U, Maiti S, Webb WW (1999) Molecular dynamics in living cells observed by fluorescence correlation spectroscopy with one- and two-photon excitation. Biophys J 77:2251–2265
22. Hess ST, GirirajanTPK, Mason MD (2006) Ultra-high resolution imaging by fluorescence photoactivation localization microscopy. Biophys J 91(11):4258–72
23. Betzig E, Patterson GH, Sougrat R et al (2006) Imaging intracellular fluorescent proteins at nanometer resolution. Science 313:1642
24. Rust MJ, Bates M, Zhuang X (2006) Sub-diffraction-limit imaging by stochastic optical reconstruction microscopy (STORM). Nat Methods 3:793
25. Curthoys NM et al (2019) Influenza Hemagglutinin Modulates Phosphatidylinositol 4,5-Bisphosphate Membrane Clustering. Biophys J 116:893–909
26. Raut P et al (2022) Phosphatidylinositol 4,5-Bisphosphate mediates the co-distribution of influenza a hemagglutinin and matrix protein M1 at the plasma membrane. Viruses 14(11):2509
27. Varghese JM et al (2023) Single-Molecule Super-Resolution Imaging Reveals Formation of NS2B3 Protein Clusters on Mitochondrial Network Leading to its Fragmentation during the Onset of Dengue (Denv-2) Viral Infection, BioRxiv. https://doi.org/10.1101/2022.12.18.520514
28. Varghese JM (2023) The Anti-inflammatory Drug Leflunomide Inhibits NS2B3 Cluster Formation During Dengue Viral Infection as Revealed by Single Molecule Imaging, BioRxiv. https://doi.org/10.1101/2023.12.26.573168
29. Magde D, Elson E, Webb WW (1972) Thermodynamic fluctuations in a reacting system—measurement by fluorescence correlation spectroscopy. Phys Rev Lett 29:705
30. Elson EL, Magde D (1974) Fluorescence correlation spectroscopy. I. Conceptual Basis and Theory Biopolymers 13:1–27
31. Magde D, Elson EL, Webb WW (1974) Fluorescence correlation spectroscopy. II. An Experimental Realization Biopolymers 13:29–61
32. Zander et al (1996) Appl Phys B 63:517–523
33. Mondal PP (2022) Classical quantum optics. AIP Publishing, New York
34. de Vries H, Wiersma DA (1980) J Chem Phys 72:1851
35. Bernard J, Fleury L, Talon H, Orrit M (1993) J Chem Phys 98:850
36. Ambrose WP, Basché T, Moerner WE (1991) Detection and spectroscopy of single pentacene molecules in a p-terphenyl crystal by means of fluorescence excitation. J Chem Phys 95:7150–7163
37. Patterson FG, Lee HWH, Wilsan WL, Fayer MD (1984) Chem Phys 84:51
38. de Vries H, Wiersma DA (1979) J Chem Phys 70:5807
39. Saep B, Kellrnann A, Martin M, Lindqvist L (1972) Chem Phys Lett 13:241
40. Kummer S, BaschC T, Brauchle C (1994) Chem Phys Lett 229:309; Chem Phys Lett 232:414 (1995)

41. Basche T et al (1996) Single molecule optical detection, imaging, and spectroscopy. VCH Publication, Weinheim
42. Kummer S, BaschC T, Brauchle C (1994) Clzem Phys Lett 229:309; Chem Phys Lett 232:414 (1995)
43. Moerner WE, Kador L 1989 Optical detection and spectroscopy of single molecules in a solid. Phys Rev Lett 62:2535–2538
44. Dehmelt HG (1975) Bull Am Phys SOC2 0:60
45. Scully MO, Zubairy MS (1997) Quantum optics. Cambridge University, Cambridge
46. Nagoumey W, Sandberg J, Dehmelt H (1986) Phys Rev Lett 56:2797
47. Sauter T, Neuhauser W, Blatt R, Toschek PE (1986) Phys Rev Lett 57:1696
48. Bergquist JC, Ilulet RG, Itano WM, Wineland DJ (1986) Phys Rev Lett 57:1699
49. Kummer S, BaschC T, Brauchle C (1994) Clzem Phys Lett 229:309; Chem Phys Lett 232:414(1995)
50. Loudon R (1983) The quantum theory of light. Oxford University, Oxford
51. Brown R, Wrachtrup J, Orrit M, Bernard J, von Borczyskowski C (1994) J Chenz Phys 100:7182
52. Mondal PP (2022) Classical and quantum optics. AIP Publishing, USA
53. Basche T, Moerner WE, Orrit M, Talon H (1992) Photon antihunching in the fluorescence of a single dye molecule trapped in a solid. Phys Rev Lett 69:1516
54. Carmichael HJ, Walls DF (1976) J Phys B 9:L43
55. Kimble HJ, Dagenais M, Mandel L (1977) Phys Rev Lett 39:691
56. Diedrich F, Walther H (1987) Phys Rev Lett 58:203
57. Bernard J, Fleury L, Talon H, Orrit M (1993) Photon bunching in the fluorescence from single molecules: a probe for intersystem crossing. J Chem Phys 98:850

Optical Design and Instrumentation of Single Molecule Optical Microscope (fPALM/PALM/STORM)

9

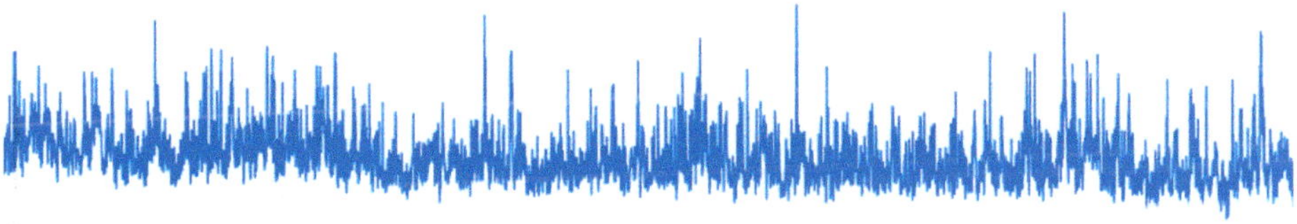

Among other optical microscopy systems, the optical design and instrumentation behind single-molecule localization microscopy (SMLM) are relatively complex and calls for specialized skill. However, SMLM comes with the distinct ability to resolve features well beyond classical diffraction limit of $\lambda/2$, thereby cause a PArticle Resolution shift (PAR-shift) towards the single-molecule limit. Foremost, the techniques demand precision illumination and ultra-sensitive detection of the Stoke-shifted photons from activated molecules. The fact that molecules need to be prepared in a fluorescence state by light-induced photochemical reactions through the photoconversion process calls for additional light sources. This makes the microscope a dual-beam illumination system. In this chapter, we will discuss three different illumination schemes that are widely used in single-molecule localization microscopy: widefield, TIRF, and HILO. Most of the popular techniques (fPALM/PALM/STORM and variants) use one of these optical configurations for exciting the sample. In addition, recent progress in SMLM has seen the emergence of volume imaging techniques (3D STORM, fPALM-Biplane, SMILE, and *scanSMLM*) and a steady advance towards multicolor imaging, thereby facilitating the investigation of multiple proteins interactions. All these developments are possible with the emergence of single-photon sensitive detectors such as EMCCD, sCMOS, and SPAD. The final section of the chapter is dedicated to these sophisticated detectors and their working principle. The chapter is expected to give first-hand knowledge for the development of basic SMLM system.

P. P. Mondal, S. Hess, *Single-Molecule Super-Resolution Optical Microscopy*,
https://doi.org/10.1007/978-3-032-13442-4_9

Right from the inception of the idea of resolution in the early 1900s, it was believed that the resolving power is limited by the diffraction of light and features finer than this limit cannot be resolved using visible light [1–3]. However, in the year 1952, Toraldo di Francia proposed pupil-plane filtering to increase system resolution [4]. The technique uses a set of concentric amplitude and/or phase filters at the pupil of the imaging lens. This, along with wavefront engineering and computational deconvolution, showed the focusing of a spot beyond the Abbe-Rayleigh diffraction limit [5–8]. The quest for super-resolution was again taken up after four decades when, in early 1990s, Stephen Hell proposed a new concept based on stimulated depletion, giving rise to STimulated Emission Depletion (STED) microscopy [9, 10]. STED falls under RESOLFT concept where temporal ON/OFF switching of selective nearby molecules (well within the diffraction limit) is carried out to achieve very high resolution (<10 nm). This technique comes with the distinct advantage of super-resolution (resolution much below the diffraction limit). We will discuss STED microscopy in Chap. 10. Subsequently, the year 2005–2006 saw a break-through in super-resolution with the emergence of two potential new concepts: structured illumination and single molecule based SMLM (fPALM/PALM/STORM). While structured illumination uses spatially structured pattern light to enhance resolution two-fold as compared to standard widefield microscopy, SMLM uses a combination of single molecule PSFs and computational localization methods to improve resolution far beyond the diffraction limit (tens of nanometers or better). This is incredible considering the advantages of SMLM techniques, i.e., low-intensity illumination, sparsity, and independent detection of single molecule PSFs. These features make it very suitable for imaging delicate biological specimens (such as, live cell) that are prone to photobleaching and phototoxicity. However, SMLM techniques require acquisition of multiple images for reconstructing a single super-resolved image that compromises temporal resolution [11].

Among all the available single molecule based super-resolution techniques, SMLM has seen wide popularity in recent times. In general, SMLM can be realized on three major optical configurations (widefield, TIRF, HILO). Apart from that, the detection may be single or multi- channel, depending on the application. Of special attention are the different kinds of sophisticated detectors (EMCCD, sCMOS, SPAD) that are used in modern SMLM implementations. In this chapter, we will discuss the basics of optical design for realizing a complete SMLM system that makes use of these specialized ultrasensitive detectors.

9.1 Illumination Schemes

Super-resolution microscopy have three major optical illumination schemes: Widefield, Total internal reflection and HILO. All these schemes are widely used and have their advantages. While widefield may enable examining the entire 3D specimen, total internal reflection can interrogate very thin slice of specimen (about 100 nm from the coverslip), and HILO allows sectional imaging far from coverslip. Here we will discuss all the schemes in detail.

9.1.1 Widefield Illumination

This is the most common mode used in optical microscopy where the specimen is exposed to a wide light field (circular intensity profile followed by cropping the field-of-view to keep only the uniform central area) illuminating the entire specimen. A large fraction of latest widefield epifluorescence microscope use Kohler illumination (see, Fig. 9.1a). In Kohler illumination, the light from different points (say, P1 and P2) of the light source (filament) were mixed by the collector lens and thus scrambles the light source structure (filament) (see, Fig. 9.1b). Note the blue and black rays that illustrate the path of light originating from two different points (P1 and P2) on the light source (filament). This means that the illumination is highly uniform and free from light source structures (filament). Specifically, note that image of light source does not form on the sample and detector/camera plane (see, orange lines in Fig. 9.1b). Kohler illumination method ensures

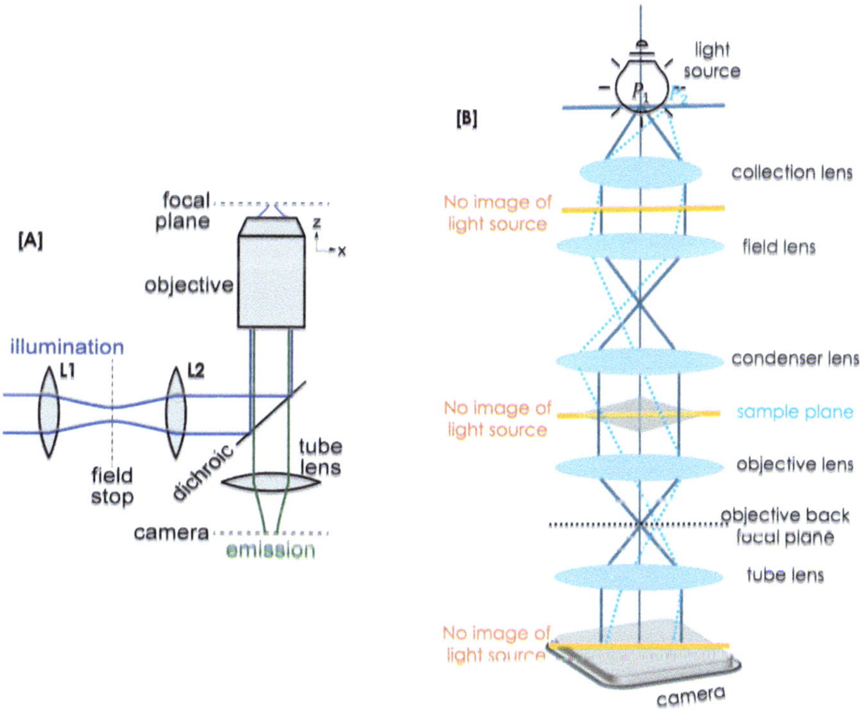

Fig. 9.1 (**a**) A typical configuration of a widefield optical microscopy. (**b**) An optical setup using Kohler illumination. The figure is adapted from, Y. L. Wang et al., Confocal imaging capacity on a widefield microscope using a spatial light modulator. PLoS ONE 16, e0244034 (2021)

that the visible structures (filament) are no longer recognizable and hence the image is free from undesirable artifacts.

In widefield configuration, Kohler illuminated light is focused onto the sample by an objective, and the same objective is used to capture emission from the sample (see, Fig. 9.1a). The emitted light is focused by a tube lens onto a camera/detector to form the image. Note that, additional optics are often used in the illumination beam path (represented by lenses L1 and L2) to shape and scale the illumination light. These optics create two separate planes, the field stop and aperture stop (not shown) where masks can be used to manipulate the excitation. The basic optical arrangement and its simplicity makes widefield microscope accessible to every research laboratory.

It may be noted that, Kohler illumination is necessary for light emitted by a large light source emitting non-uniform light (unlike laser), and the one that have source structure. Some of the examples are Halogen/neon lamps with a filamentous structure.

9.1.2 TIRF Illumination

Although the widefield detection scheme is simple and effective, it suffers from light coming from nearby specimen planes/off-focal planes of the object. This affects the overall quality (contrast, SNR and to some extent resolution) of the reconstructed image. To overcome this potential issue, total internal reflection optical mode can be used. However, the imaging can be performed only near to the surface of coverslip (upto a depth of few hundred microns).

Figure 9.2 shows a schematic diagram of a typical optical configuration used in TIRF microscopy. The detection technique is similar to that of widefield super-resolution microscopy, except for the illumination sub-system. A thin pencil of incident beam is directed to the back-aperture of an objective lens by a combination of optics. The optics is used to focus the light to one end of the objective back-aperture. This bends the incident light greater than critical angle (θ_C), resulting in total internal reflection at the coverslip-

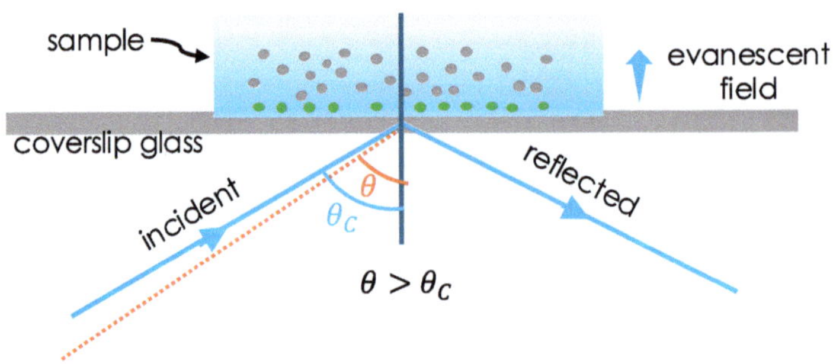

Fig. 9.2 TIRF based single molecule based super-resolution system. The figure is inspired from Partha P. Mondal, Samuel T. Hess, Total internal reflection fluorescence based multiplane localization microscopy enables super-resolved volume imaging, Appl. Phys. Lett. 110, 211102 (2017)

sample interface. As a consequence, an evanescent field builds us near the coverslip, with field strength falling rapidly with distance (far from coverslip) as shown in Fig. 9.2. The field and the penetration depths (γ) of the evanescent field are given by [12],

$$|\vec{E}|^2 = |E_0|^2\, e^{-z/\gamma} \tag{9.1}$$

where, the penetration depth is,

$$\gamma = \lambda_0/2\pi n_2\sqrt{[sin\theta_i/sin\theta_c]^2 - 1} \tag{9.2}$$

and the critical angle is,

$$\theta_c = sin^{-1}(n_2/n_1) \tag{9.3}$$

For a typical setup, $n_1 = 1.51$ (refractive index of coverslip glass), $n_2 = 1.33$ (refractive index of specimen) and $\lambda_0 = 561$ nm (wavelength of incident light), gives a critical angle of $\theta_c = 61.738°$ and a penetration depths of, $\gamma \approx 1400$ nm.

9.1.3 HILO Illumination

Highly Inclined and Laminated Optical-sheet (HILO) is a technique that is somewhat between widefield and TIRF and hence benefits from the advantage of both the illumination schemes. In this configuration, the laser is directed at sharp angle through the sample illuminating a small section of the specimen. Unlike TIRF configuration, the incident angles are always less than critical angle i.e, $\theta < \theta_c$. This allows large penetration depths, however with a signal-to-background noise ratio (SNR) slightly lower than that of TIRF. This is also the reason why it referred to as "Dirty TIRF".

A classic configuration of HILO microscopy is shown in Fig. 9.3b. Alongside, its comparison with widefield/epifluorescence optical configuration is also shown (Fig. 9.3a). HILO involves inclining the illumination beam by a large refraction and is laminated as a thin optical-sheet in the specimen plane (just over the coverslip). This is technically achieved by moving the condenser lens along with the incoming incident beam (using, a mirror-lens combination on a single movable platform). This thin optical sheet is employed for illuminating the specimen, and infact many such layers of the specimen can be imaged. The thickness (Δz) of the field can be approximated by [13],

$$\Delta z = R/\tan\theta \tag{9.4}$$

where R is the diameter of the illuminated area and θ is the angle at the specimen plane (see, Fig. 9.3). Note that, the estimation of field thickness ignores divergence.

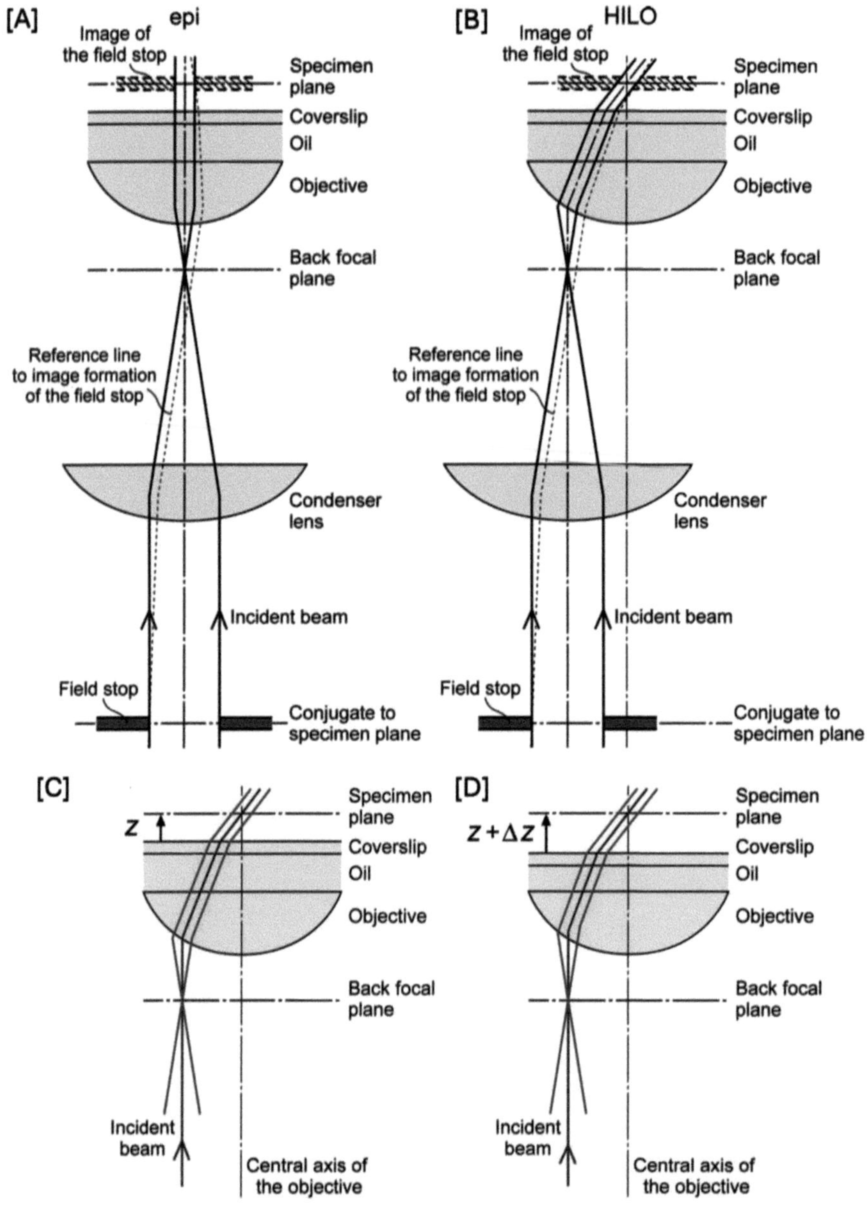

Fig. 9.3 The beam-paths for, (**a**) Epi illumination, and (**b**) Highly Inclined and Laminated Optical-sheet (HILO) illumination. HILO allows change of beam-path with incident angle (θ) that allows change in ΔZ. The figure is adapted from, M. Tokunaga et al., Nat. Meth. 5, 159–161 (2008)

9.2 Single Molecule Localization Microscopy

Single molecule localization microscopy (SMLM) is probably the most powerful microscopy technique developed so far for studying live biological specimens. This has the power to reveal the ongoing biological processes at truly single molecule level. Below we discuss few key SMLM techniques.

9.2.1 Fluorescence Photo-Activation Localization Microscopy (PALM/fPALM)

As the name suggests, fPALM/PALM are a class of techniques that use lights of two distinct frequencies to activate and excite the target fluorophore. However, exceptions do exist. The microscopy technique can be realized in two optical configurations (Widefield and TIRF) as discussed below:

9.2.2 Widefield Super-resolution Microscopy

To realize widefield super-resolution microscopy, a typical optical microscopy system is as shown in Fig. 9.4. The technique uses two frequencies of light, one for readout and a second one for activation. The beams from a collimated light source (preferably a laser) are combined by a dichroic mirror (DM1) and expanded using a beam-expander system

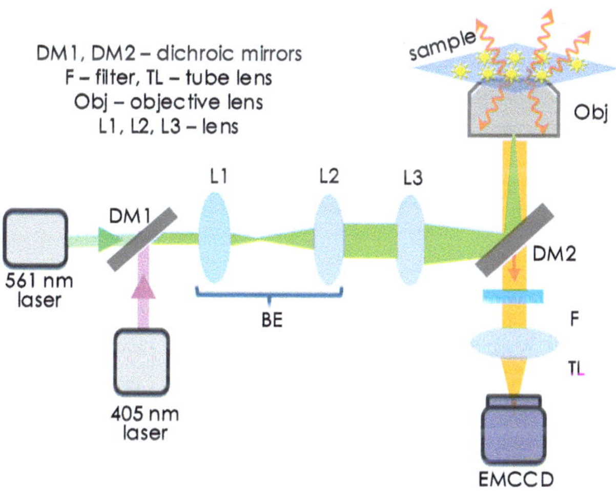

Fig. 9.4 The first full-fledged single molecule based super-resolution system. The figure is inspired from S. Hess et al., Biophys. J. 19, 4258–4272 (2006)

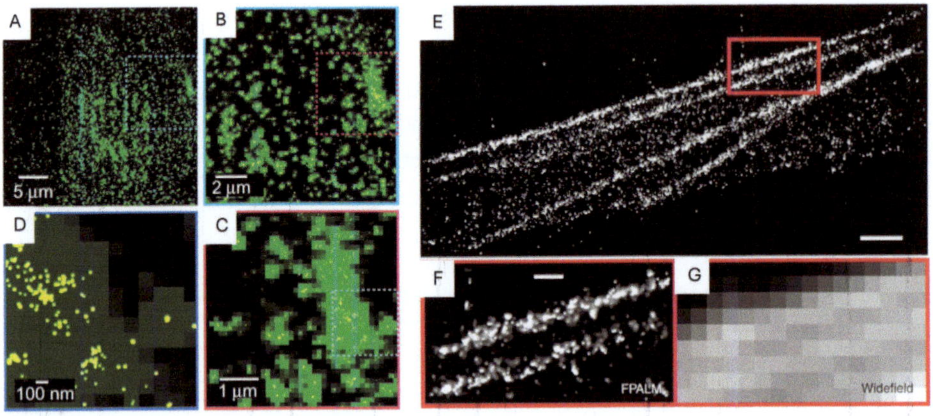

Fig. 9.5 (**a–d**) FPALM images of PA-GFP on a glass coverslip, with 48,746 localized molecules (in yellow) obtained from Gaussian fit. The sequence from **a** to **d** progresses toward higher zoom. (**e, f**) Actin filaments in a Dendra2-actin expressed in a fixed mouse fibroblast cell as resolved by fPALM microscopy. (**g**) Conventional widefield fluorescence image of Actin filaments. Images shown in (E-G) and (A-D) are taken from, T. J. Gould et al., Nature Protocols 4, 291 (2009) and S. T. Hess et al., Biophys J. 91, 4258–4272 (2006), respectively

(BE). Expansion is achieved by a combination of lenses, say two biconvex lenses (L1 and L2) of different focal lengths. Subsequently, the beams are focused by a lens L3 to the back of a high NA objective lens. Note that, the lens L3 is located in the rear port of an inverted microscope exactly at one focal length from the back-aperture of the objective lens (obj). This produces a uniform widefield illumination over the covership that holds the specimen. Within the region illuminated by activation beam (405 nm), the inactive molecules are activated which are then excited by another beam (excitation light of wavelength 561 nm). Note that only the activated molecules undergo excitation. The fluorescence emitted by the activated molecules are then collected by the objective and directed to the dichroic mirror (DM2) which allows it to pass through to the detector. On the way, the light is filtered by a series of optical filters (F) to eliminate undesired background, illumination beams, and other radiations. A tube lens (TL) is used to focus the filtered fluorescence on to the detector chip of a sensitive EMCCD detector. The image thus formed in the detector are that of activated molecules/point sources that appear as bright spots in the image plane. This is followed by determining the background and quantifying the spot to estimate the number of emitted photons. Figure 9.5 shows the fPALM images PAGFP protein spread on a coverslip, and Actin filaments in a cell.

9.2.3 TIRF Based Super-resolution Microscopy

Figure 9.6 shows a schematic diagram of a typical optical setup used in TIRF based super-resolution microscopy. The technique is similar to that of widefield super-resolution

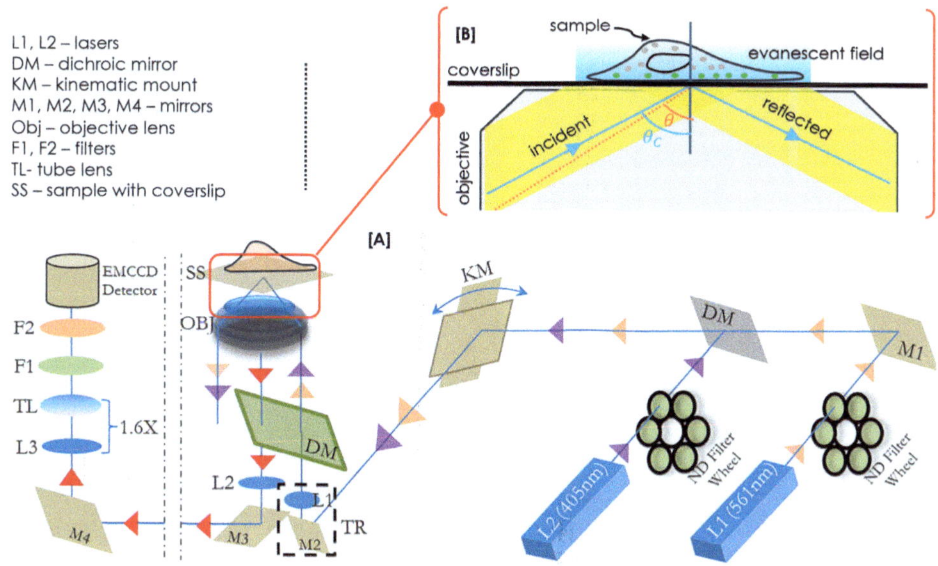

Fig. 9.6 TIRF based single molecule based super-resolution system. The figure is inspired from Partha P. Mondal, Samuel T. Hess, Total internal reflection fluorescence based multiplane localization microscopy enables super-resolved volume imaging, Appl. Phys. Lett. 110, 211102 (2017)

microscopy, except for the illumination sub-system. As usual, two light sources were used for activation and excitation of photoactivable molecules used. The intensities of both the beams are controlled by neutral density filter wheels (ND). Both the beams are combined by a dichroic mirror (DM) and directed to the back-aperture of an objective lens (housed in an inverted microscope) by a combination of kinematic mirror (KM) and lens (L1). The combination is used to focus the light to the periphery (one end) of the objective back-aperture. This allows bending of incident light greater than critical angle (θ_C), resulting in total internal reflection at the coverslip-sample interface. As a consequence, an evanescent field builds us near the coverslip, with field strength falling rapidly with distance (far from coverslip) as shown in Fig. 9.6b. The detection sub-system is similar to that of widefield configuration, with the exception of additional magnification (here, $1.6X$). The field and the penetration depths (γ) of the evanescent field are given by Eqs. (9.1–9.3) [12]. Figure 9.7(A,B) shows widefield, and TIRF images of fluorescent beads and actin filaments, whereas (C,D) are TIRF and TIRF-SMLM images of cytoskeleton and actin in a cell, respectively. However, Fig. 9.7(A-F bottom-half) images are that of TIRF-widefield and TIRF-PALM images of molecular clusters.

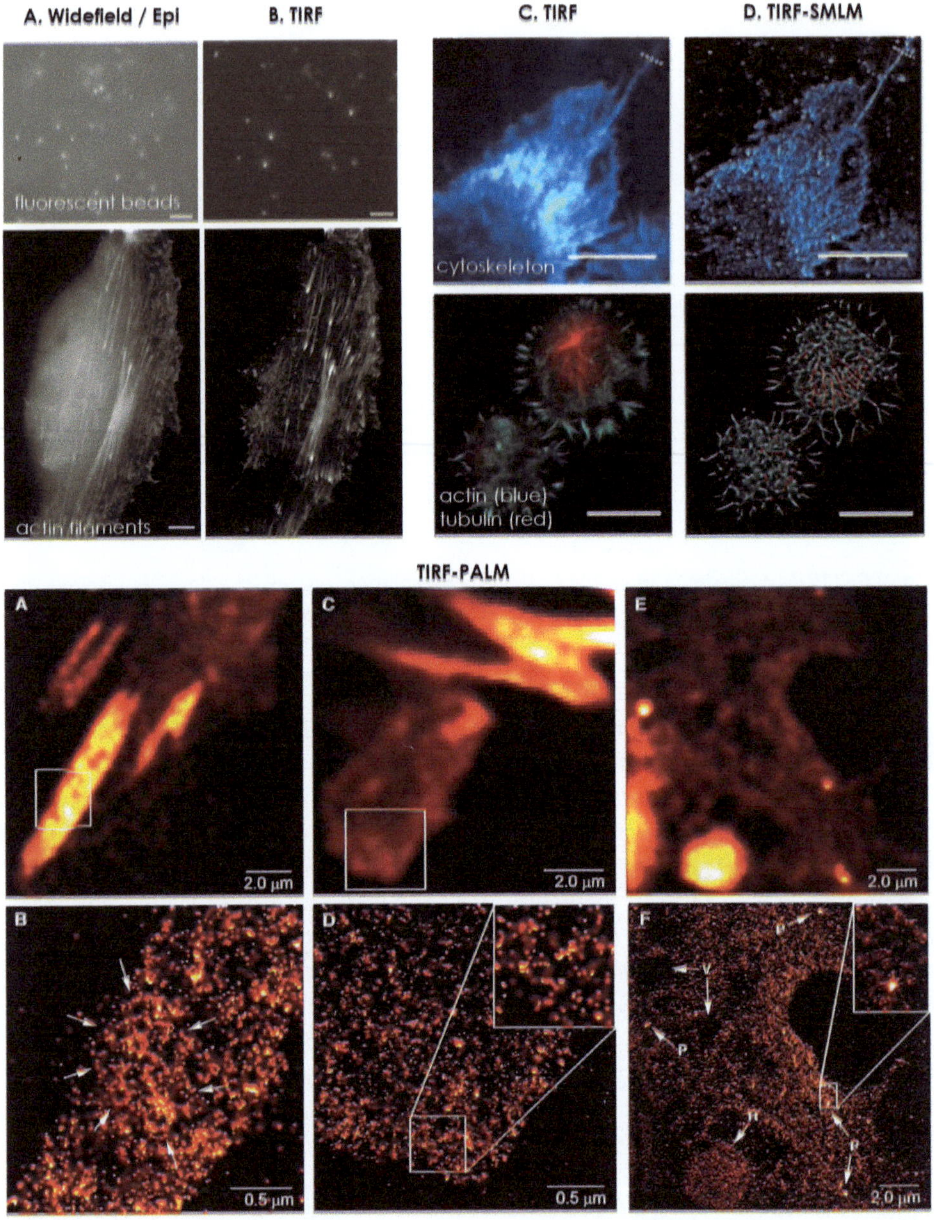

Fig. 9.7 TIRF based single molecule localization microscopy images of fluorescent beads, Actin, Tubulin and single molecule clusters. Images are taken from, J. Biophotonics 9, 948-957 (2016); and, E. Betzig et al., Imaging intracellular fluorescent proteins at nanometer resolution, Science 313, 1642–1645 (2006)

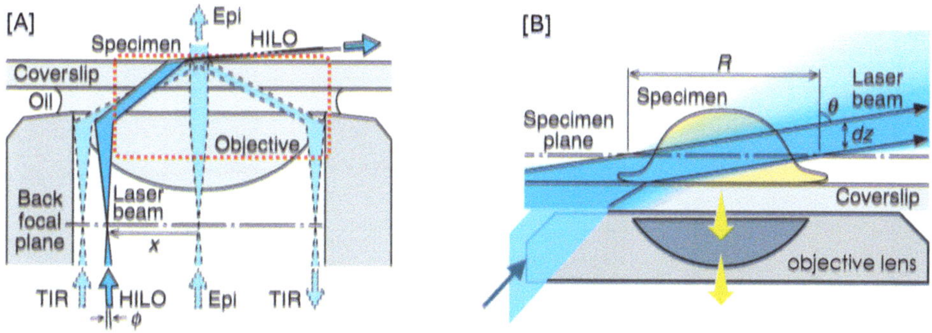

Fig. 9.8 Highly Inclined and Laminated Optical-sheet (HILO) based single molecule based super-resolution system. The figure is inspired from, M. Tokunaga et al., Nat. Meth. 5, 159–161 (2008)

9.2.4 HILO Based Super-Resolution Microscopy

A typical HILO based super-resolution microscopy system is schematically shown in Fig. 9.8. In this optical configuration, the incident laser laser bean is highly inclined by a large refractive index objective lens. Practically, this is obtained by focusing light from the near-periphery of the back-aperture of the objective lens such that it is between widefield and TIRF illumination configuration. This allows a thin optical sheet at the specimen side with the z-sectioning decided by the incident angle at the specimen. Few celebrated images of nuclear pore complexes in a 3D cell at varying depths, fluorescent beads and actin-cytoskeleton images are shown in Fig. 9.9(a–e).

9.2.5 STochastic Optical Reconstruction Microscopy (STORM)

The optical configuration of STORM and other SMLM techniques (such as, PALM and fPALM) are identical. All these techniques require two different frequencies of light, one for photoactivation/photoswitching/ photoconversion of single molecules and the other for excitation of activated molecules. STORM microscopy can be realized in all the three optical configurations (widefield, TIRF and HILO). The technique involves imaging photoswitchable dyes (e.g., Cy3), that are bright and photostable over a large number of emission cycles.

A STORM system has an optical configuration similar to that of PALM/fPALM system as discussed in the previous sections. Typically, STORM imaging may be performed on a commertial optical microscopes with lasers for photo-switching and excitation of single molecules. As an example, a 532 nm laser is used for photo-switching of Alexa 647 dye molecules and 657 nm for its excitation. As usual, two lasers are combined by a beam-combiner (dichroic mirror). Subsequently, the combined beams are either directly focused or coupled to an optical fiber, followed by collimation and focussing on to the back-focal

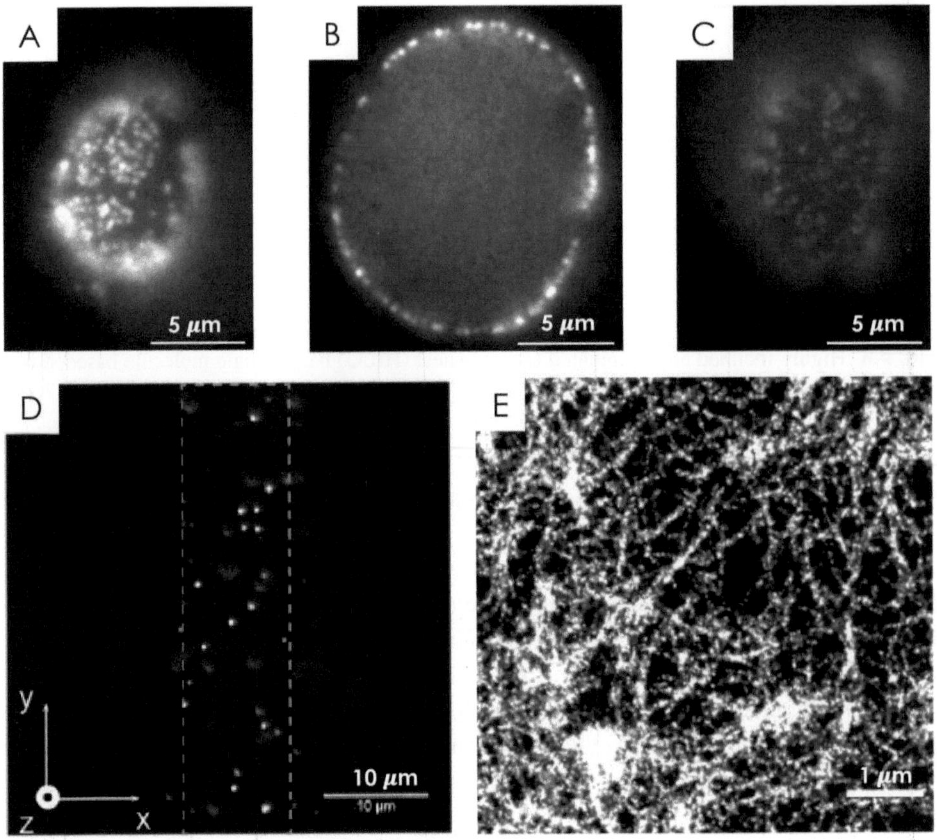

Fig. 9.9 HILO based imaging of nuclear pore complexes in a 3D cell at depths of, (**a**) 0.9 μm, (**b**) 5.7 μm, and (**c**) 14.7 μm from the coverslip. (**d**) Images of 100 nm fluorescent beads at a beam inclination of $\theta = 77°$. (**e**) Single molecule based super-resolved images of actin cytoskeleton labelled with Alexa Fluor dye ($AF488$). The images are taken from, M. Tokunaga et al., Nat. Meth. 5, 159–161 (2008) ; L. Gardini at al., Optics Exp. 31, 26208 (2023)

plane of a high NA oil immersion objective through the back port of the microscope. A translation stage can be employed for shifting both the beams towards the edge or center of the objective lens so that STORM imaging can be performed in widefield/ TIRF/HILO mode. Subsequently, the emission is collected by the same objective and filtered by combination of band-pass and notch filters. The filtered emission is then directed onto the EMCCD camera through a tube lens or through a pair of lenses for further optical magnification.

During data collection, single fluorescent molecules are switched ON/OFF using different colors of light, for localizing their positions with high accuracy. A series of imaging cycles are considered to record sufficient single molecule PSFs necessary to construct super-resolved map of sub-cellular structures. In each imaging cycle, a fraction of single molecules are switched on by a specific frequency of light (activation light) so

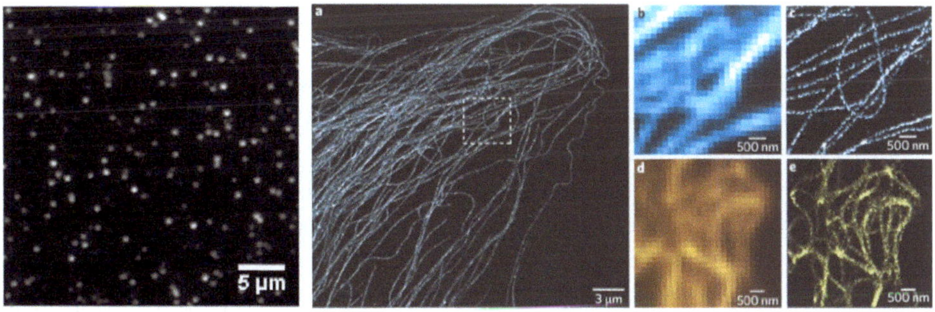

Fig. 9.10 STORM imaging of Cy5 molecules and microtubules (labeled with Alexa Fluor 647). The images are taken from, Nature Photonics 3, 365–367 (2009)

as to avoid overlapping of PSFs. This ensures that the molecules are optically resolvable. Such a process allows the position of these fluorescent molecules to be determined with high precision (see, next section for localization precision). In each cycle, a small subset of molecules are photoconverted from non-fluorescent to fluorescent state i.e, turned-ON and further excited to record its PSF image (as a bright spot). This allows the position to be determined with high accuracy. Repeating many such cycles gives the position of several single molecules which can be put-together to reconstruct an overall map of the molecule distribution. The collective distribution of these targeted molecules may ultimately represent structure of organelles (such as, Actin/Mitochondriam/Golgi-bodies etc.) or a specific biological activity (protein clustering in Influenza, Dengue, Ebola etc). STORM is similar to a widefield microscopy in terms of visualizing structures but with sub-diffraction resolution and single molecule precision. Fig. 9.10 shows the images of fluorescent dyes (Cy5) and microtubules in a cell. The microtubules are immunostained with antibodies that are labeled with photo-switchable Alexa Fluor 647.

9.3 Localization Precision of Single Molecules

The resolution of a light microscope is $r_0 \sim 0.61\lambda/NA \sim \lambda/2$, with λ as the wavelength of light. So, any object $<\lambda/2$ will appear as diffraction limited spot (with a size of $\sim\lambda/2$). However, the center of the spot i.e, its location can be determined with much greater precision. This allows the determination of relative positions of two objects with a precision much better than $\lambda/2$. This is valided in a number of experiments ranging from particle tracking to molecular motions [14–18].

The localization precision of an object depends on a number of factors, of which most prominent ones are, (1) shot noise of photons in the spot, (2) pixelation noise, and background noise. In short, the photon shot noise occurs when the noise in each pixel is dominated by the limited number of photons detected from the particle / molecule itself. The background noise is the noise due to other sources such as, dark current noise, readout noise, autofluorescence, out-of-focus molecules and other sources of fluorescence. However, the pixelation noise is solely due to the finite size of the pixel that gives rise to the uncertainty of the photon location within the pixel. We will discuss all the sources of noise and derive a term for the best estimate for the position. Finally, all the terms will be put together which will give us the resultant estimate [18].

First, we take up the case of short noise or photon-noise limited case. In this case, each photon emitted by the object gives a measure of the position of the object, and the associated position error of each measurement (standard deviation of PSF). The estimate of position (variance) is essentially the average of all positions with error as the standard error i.e.,

$$\langle(\Delta x)^2\rangle = s^2/N \tag{9.5}$$

where, s is the standard deviation of PSF, and N is the number of measurements (here, number of photons) or sample size.

The pixelation noise is due to the finite size of the pixel/sensor in the camera chip. This arise due to the uncertainty of where the photon will fall on the square pixel causing variation limited by its size. This results in flat-top or top-hat distribution of size (a) of the pixel for $s > a$. This variance for top-hat distribution is, $a^2/12$. The corresponding uncertainty or variance is given by,

$$\langle(\Delta x)^2\rangle = \frac{a^2/12}{N} \tag{9.6}$$

Finally, we determine the expression for the background noise caused by other sources (other than the object). A straightforward way is to determine the least squared error between the actual pixel count (say, y_i) and expected number of count from the object located at x with σ_i as the expected count uncertainties. Here, the uncertainties due to photon-counting noise (which is embedded in σ_i) is,

$$\sigma_i^2 = N_i(x) + b^2 \tag{9.7}$$

where, $N_i(x)$ and b are sum of photon counting noise and background noise, respectively.

The least squared fit is given by,

$$\chi^2(x) = \sum_i \frac{(y_i - N_i(x))^2}{\sigma_i^2} \tag{9.8}$$

This needs to be minimized i.e,

$$\frac{d\chi^2}{dx} = 0 \tag{9.9}$$

which gives an expression for measured position x in terms of counted photons in each bin.

For the case of pure background noise (i.e, without photon noise), N_i can be approximated as [18],

$$N_i = \frac{N}{\sqrt{2\pi}s} e^{-i^2/2s^2} \tag{9.10}$$

and the variance is given by,

$$\langle(\Delta x)^2\rangle = \frac{4\sqrt{\pi}s^3b^2}{aN^2} \tag{9.11}$$

Combining the above set of equations (Eqs. 9.5, 9.6, and 9.11), we get,

$$\langle(\Delta x)^2\rangle = \frac{s^2 + a^2/12}{N} + \frac{4\sqrt{\pi}s^3b^2}{aN^2} \tag{9.12}$$

Till now, the derivation is done for one dimension. Now considering two-dimension for 2D PSF, the background noise need to be calculated over two-dimensional integral. This gives the resultant variance as [18],

$$\langle(\Delta x)^2\rangle = \frac{s^2 + a^2/12}{N} + \frac{8\pi s^4b^2}{a^2N^2} \tag{9.13}$$

The above Eq. (9.13) represents the complete expression for the localization precision of a fluorescing single molecule. In SMLM, the accuracy of particles position (localization precision) in a noisy condition can be determined based on maximum likelihood estimation (MLE) [19, 20]. The expression (Eq. 9.13) will be used again and again in the subsequent sections to determine the location and size of single molecules.

9.4 3D SMLM Techniques

Three dimensional SMLM is a powerful technique to realize the structure of sub-cellular organelles. Here, we will discuss a few handful techniques and explore their benefit towards volume super-resolution imaging.

9.4.1 Astigmatic 3D STORM Imaging

Three dimensional STORM uses astigmatism to realize 3D structures of sub-cellular entities. This stems from the fact that astigmatism induces artefacts to a point emitter (e.g., single molecule), thereby discerning their position dependent shape (in the image plane) revealing its location in the 3D specimen.

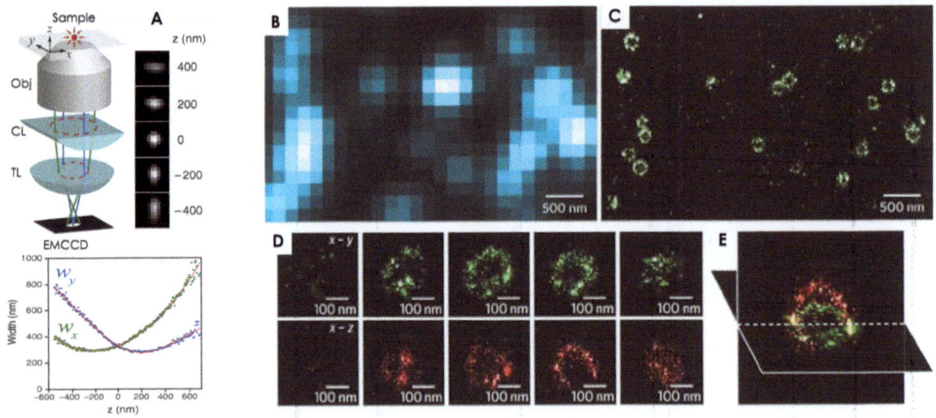

Fig. 9.11 (**a**) The detection sub-system of a 3D STORM system. Alongside calibration plot of image -widths w_x and w_y as a function of z for the single molecule is also shown. Note that illumination sub-system is similar to that of a typical SMLM system. (**b**) Conventional widefield image of clathrin-coated pits (CCP) in a region of cell (XY cross-section). (**c**) 3D STORM image of the same (XY cross-section). (**d**) XY and XZ cross-sections of CCPs. (**e**) 3D STORM image of a single CCP. Images taken from, B. Huang, W. Wang, M. Bates, X Zhuang et al., Science 319, 810–813 (2008)

In this approach, the illumination sub-system remains similar to that of widefield SMLM while the detection sub-system is different as shown in Fig. 9.11. As usual, the system is a standard $4f$ system comprising objective lens (Obj) and tube lens (TL). For the astigmatism imaging method, a weak cylindrical lens (focal-length of 1000 cm) is introduced in the Fourier plane of the objective lens i.e, between objective lens and tube lens. This introduces two slightly different focal planes along the lateral axes x and y. As a consequence, the ellipticity and orientation of the fluorophore's image/PSF varies as its position changes in the axial z-direction (see, Fig. 9.11a). When the single molecule is in the exact focal plane of the objective, the image appeared round with equal widths in the x and y directions. If the molecule is slightly above the exact focal plane, the PSF image appeared ellipsoidal with its long axis along x-axis, while long axis is along y-axis when the molecule is below the focal plane. The calibration requires PSF image to be fitted to Gaussians with different widths in x and y (one for x and another for y) to determine the x and y coordinates of the peak position, and to obtain corresponding widths w_x and w_y which can be used to determine the z-coordinate of the fluorescent molecule (see, Fig. 9.11a). The process is repeated for every single molecule with an estimated $\geq$15,000 molecules to generate the 3D STORM super-resolved image.

Largely, three-dimensional STORM is beneficial for imaging nanoscopic sub-cellular structures. For example, the structure of spherical cage like structures of clathrin-coated pits (CCPs) in BS-C-1 cells can be images as shown in Fig. 9.11b–e [21]. These structures are of the size of $\sim$150 nm that facilitate endocytosis. To image CCPs, immunofluorescence protocols (using primary antibodies with clathrin) are employed and doubly labelled with Alexa 647 and Cy3. Both 2D and 3D STORM images are displayed in Fig. 9.11d, e that clearly show the round shape of CCPs with a size of 200 nm. Specifically, the 3D STORM image allow visualization of the CCPs in 3D (see, Fig. 9.11e). In comparison, conventional fluorescence microscopy shows a diffraction-limited blurry spot with no distinguishable structure (Fig. 9.11b). Thus 3D STORM could unambiguously resolve the ring-like structure of CPPs.

9.4.2 fPALM Biplane Imaging

As the name suggests, fPALM biplane imaging combines fPALM and double-plane detection scheme. This facilitates 3D sub-diffraction resolution of single molecules in a cell volume.

The basic schematic diagram for a fPALM-biplane imaging is shown in Fig. 9.12. Similar to standard SMLM techniques, two lasers are used for illumination, of which one is for activation and other for excitation of photoactivable molecules. It is the distinct detection sub-system that enables simultaneous two-plane imaging. Post excitation, the fluorescence light is collected by the objective lens and filtered by a dichroic mirror (D). The light is then focused by the internal tube-lens and reflected by the mirror of the inverted microscope system. In general, the camera is placed at the focal plane (FP) to record the image. However, additional relay optics can be used if additional operations are desired (such as, higher magnification or spatial filtering). Here, an additional set of lenses are used to relay the beam, a bandpass filter to filter-out undesirable light and a beam-splitter to split the filtered fluorescence equally to create two separate image planes on the same CCD/EMCCD camera. These planes correspond to two different layers of the specimen/object. However, the technique reduces signal-to-noise ratio.

The imaging is carried out by moving the piezo-driven sample stage by small z-steps and simultaneously recording two planes. The process can be used to record images throughout the entire specimen along z-axis. For example, biplane imaging of 40 nm

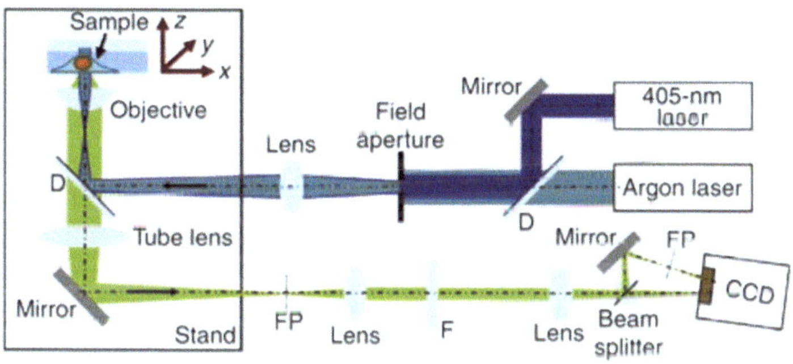

Fig. 9.12 The schematic optical setup of fPALM based biplane imaging system. The figure taken from, M. F. Juette et al., Nat. Meth. 6, 527–529 (2008)

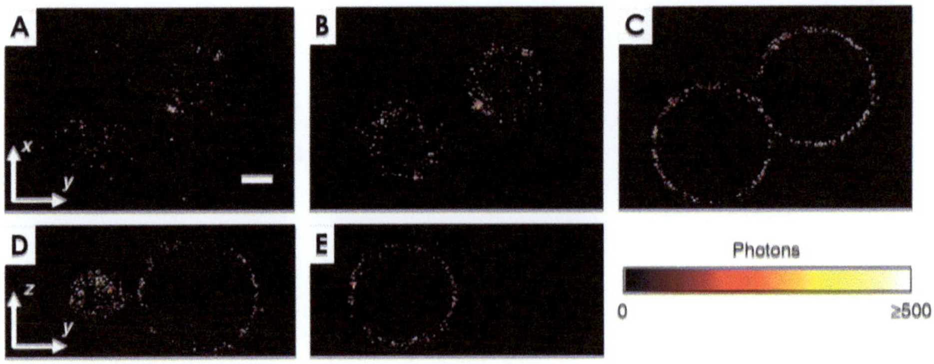

Fig. 9.13 fPALM-biplane images of fluorescent beads over a range of ~800 nm. The figure taken from, M. F. Juette et al., Nat. Meth. 6, 527–529 (2008)

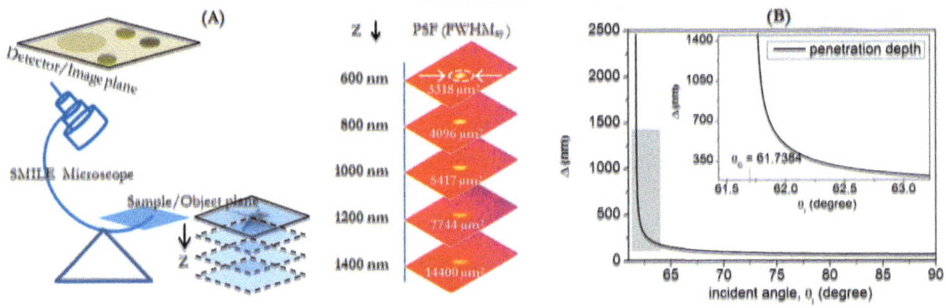

Fig. 9.14 The schematic optical setup of a TIRF based SMILE super-resolution imaging system along with the variation of PSF at varying Z-depths. Alongside, penetration depth at different incidence is also shown. The figure taken from, Partha P. Mondal, Samuel T. Hess, Appl. Phys. Lett. 110, 211102 (2017)

fluorescent beads on a coverslip is carried out by moving 100 nm z-step every 100 frames as shown in Fig. 9.13. Here the beads could be localized over a range of 800 nm. The XY and YZ sections of the reconstructed 3D volume (all localized molecules within $\pm$ 50 nm of each plane) can be seen. The technique comes with a number of advantages that include, the speed of axial scanning for volumes less than 1 micron thick, and efficient elimination of localization artefacts [22].

9.4.3 SMILE Microscopy

The SMILE super-resolution microscopy is a TIRF based 3D imaging system. The principle behind the imaging system is based on the fact that the PSF size of point emitters (single molecules) broadens due to the diffraction of light, and this can be used to reveal its z-position from the coverslip.

The working principle behind the SMILE-TIRF system is schematically shown in Fig. 9.14a. The fact that the size of single molecule PSF changes far from the coverslip-specimen interface enables 3D imaging. Fig. 9.14a shows sub-diffraction sized fluorescent beads embedded in an Agarose gel-matrix. It is apparent the beads close to the interface (i.e, at a distance of 600 nm from the coverslip) shows a diffraction-limited PSF (diameter ∼241.5 nm). This can be taken as the starting point i.e, 1st plane. As we move far from the interface (realized by z-piezo stage), the PSF size changes appreciably for single molecules located at a distance of 200 nm. Calibration with nano-beads can be carried out upto a depth of 1400 nm, beyond which the bead shape changes to non-Gaussian shape due to artefacts. The corresponding plot shows the conditions necessary ($\theta > \theta_c$) for realizing total internal reflection. The critical angle is calculated to be $\theta_c = 61.7384$ degrees. The corresponding plot gives the penetration depths versus the incident angle θ_i. Note that the penetration depth ($\gamma = \lambda_0/2\pi n_2\sqrt{(sin\theta_i/sin\theta_c)^2 - 1}$) of the evanescent field can be varied in the TIRF configuration from as low as a few tens of namometers (for $\theta_i > \theta_c$) to a couple of microns ($\theta_i \sim \theta_c$). This gives a wide range i.e, upto 1400 nm (from coverslip) for super-resolution volume imaging (Fig. 9.15).

The system is used to investigate the distribution of actin cytoskeleton in a NIH3T3-HAb2 mouse fibroblast cells. The endogeneous actin was labelled with monoclonal mouse antibody and anti-mouse secondary antibody conjugated to photoswitchable CAGE 552 dye. It is apparent that actin can form structures which span a range of length scales. The changes in the distribution of single molecules is noticable through all the 4 planes (spaced 200 nm) in the cell. It is further clear that few structures persists throughout the

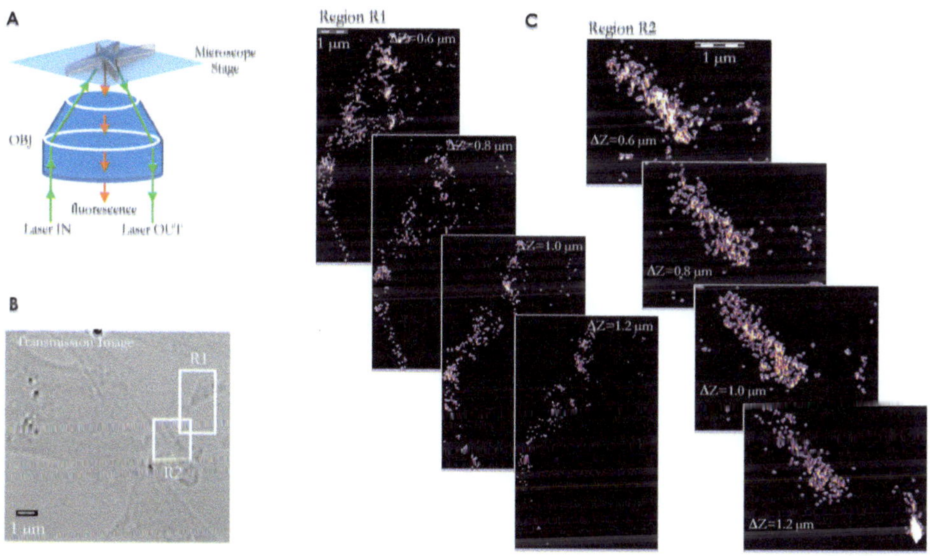

Fig. 9.15 (a–c) Single molecule super-resolved images of a cell volume stack obtained by SMILE-TIRF system along with transmission image. The figure is taken from, Partha P. Mondal, Samuel T. Hess, Appl. Phys. Lett. 110, 211102 (2017)

planes while some are restricted to a specific plane. This shows the potential of TIRF based SMILE microscopy for super-resolution volume imaging.

9.4.4 Scanning Single Molecule Localization Microscopy ($scanSMLM$)

$scanSMLM$ is an axial-scanning based super-resolution volume imaging system. Specialized optical lenses with tunable focal length (called, electrical tunable lens) is used in the detection $4f$ sub-system to enable axial scanning. When automated, the detection sub-system allows rapid scanning of axial planes of epi-illuminated object which upon stacking produces the entire super-resolved volume.

The schematic diagram of a $scanSMLM$ system is shown in Fig. 9.16. As usual, the illumination is similar to that of a standard SMLM. The fluorescence emission from the single molecules located at the focal plane is collected by the objective lens and transmitted through the dichroic mirror. The light is then reflected by the internal mirror and focused by the internal lens of an inverted microscope to the image plane. However, an additional relay optics is used that incorporates electrical tunable lens and the image forms on the camera plane which is placed at the focal length. As the name suggests, ETL is an electrically driven lens that has a fixed focal length for a specific set current value. Now by increasing/decreasing the current, the aperture of ETL changes resulting in focusing converging/diverging rays originating from off-focal planes (of the specimen/object) to the detector plane. This means that light from both focal and off-focal planes can be focused to the camera/detector rapidly and the image can be recorded. So, by changing the current one can achieve z-scanning in the object plane (see, Fig. 9.16).

$scanSMLM$ system is employed to visualize the arrangement of actin filaments and distribution of mitochondrial network in a cell volume. F-actin along with cytoskeletal polymer gives the shape to a cell and dictates mechanical properties. Hence, the arrangement, density and alignment of actin filaments are essential for chemotaxis (directed

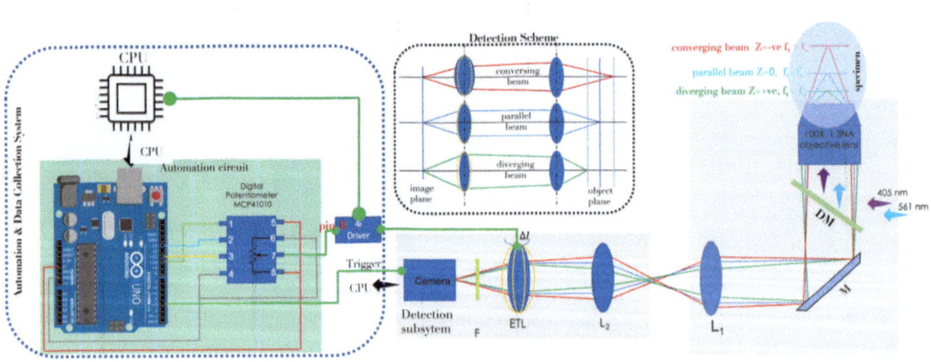

Fig. 9.16 The schematic optical setup of a scanning SMLM super-resolution imaging system. The figure is taken from, J. Basumatary, N. Baro, P. Joshi, Partha P. Mondal, Comm. Biol. 6, 1050 (2023)

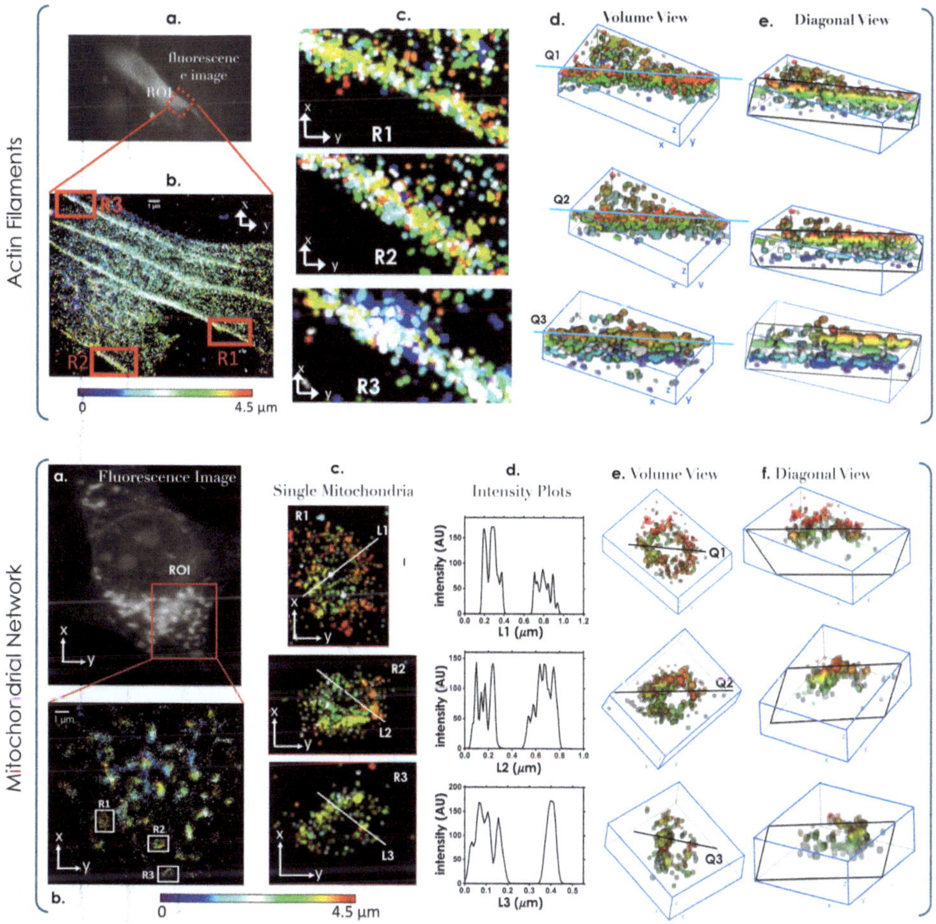

Fig. 9.17 Top block (**a–e**) and bottom block (**a–f**) are single molecule super-resolved volume image of Actin and Mitochondria in a cell volume respectively using scanSMLM system. The figures are taken from, J. Basumatary, N. Baro, P. Joshi, Partha P. Mondal, Comm. Biol. 6, 1050 (2023)

movement along a chemical gradient) and cell-cell communication. On the other hand, mitochondria known as the power-house of the cells are dynamic structures undergoing constant fission and fusion which split and merge mitochondrial tubes to form loops and branches. This forms a network which is a complex, and branched system of connected mitochondria. The network are tube-like structures with fragments ranging from 1–15 µm [23]. Visualizing these complex structures and their change over time has direct repercussions on fundamental cell and disease biology. Figure 9.17a show the scanSMLM super resolves 3D image of bundles of actin upto a depth of 4.5 µm (each slice separated by 500 nm), whereas Fig. 9.17b displays the distribution of mitochondrial network (with a clear visualization of Tom20 protein on a single mitochondria) in a cell volume. Thus *scanSMLM* 3D imaging enables direct visualization of cellular structures (actin filaments and mitochoondrial network).

9.5 Multicolor Single Molecule Localization Microscopy

The benefits of SMLM calls for its expansion in multicolor imaging. This will unleash the true potential of single molecule imaging for understanding multitude of processes simultaneously occurring in a cell, thereby expanding it reach.

A typical optical setup of multicolor fPALM imaging is shown in Fig. 9.18. Apart from all the features of standard SMLM, the multi-color variant requires the detection arm to be configured and split into two (one for each color). This is facilitated by a dichroic mirror followed by two separate filters to further clean the signals (F3 and F4 in Fig. 9.18). Both the signals (fluorescence from two different probes in the specimen) are then focused on

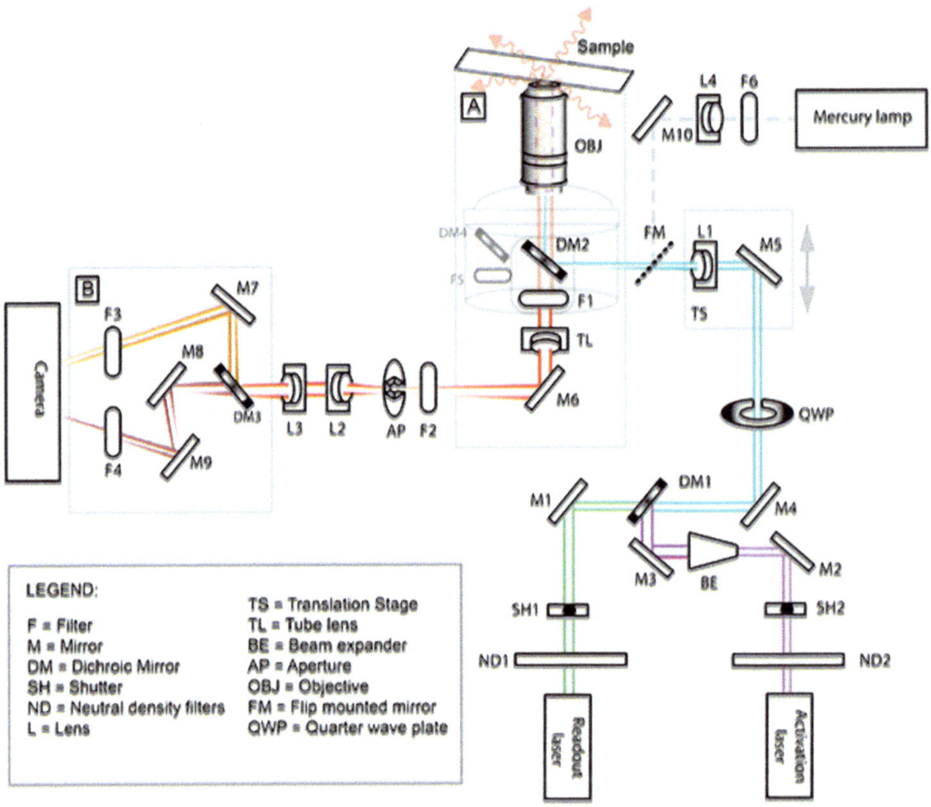

Fig. 9.18 Example schematic optical setup of a multicolor fPALM super-resolution microscopy system. The image is taken from, N. M. Curthoys et al, J. Vis. Exp. 82, 50680 (2013)

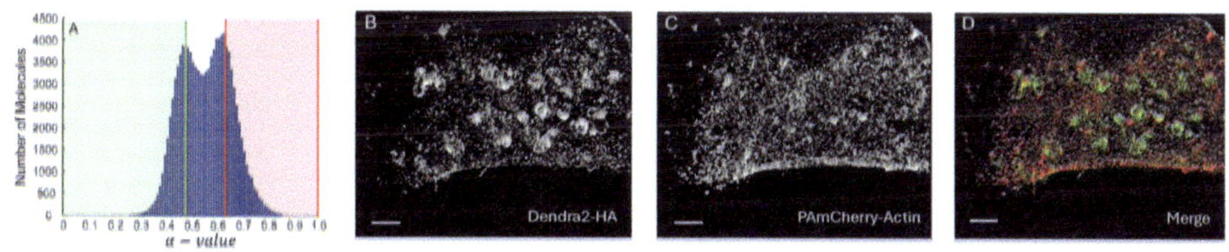

Fig. 9.19 (**a**) α histograms for the fluorophores. (**b–d**) Multicolor super-resolved images along with independent single color images. The figures are taken from, N. M. Curthoys et al, J. Vis. Exp. 82, 50680 (2013)

the camera chip (at two different locations). Using single camera to record both the signals is beneficial as this facilitates data processing and analysis.

Multicolor imaging calls for an accurate determination of α-ratio which is the ratio of the sum of the intensity of the reflected channel cutout and the sum of the intensity of both channels for each particle. The identification of the type of each molecule is carried out by comparing the relative intensities in both the channels. As an example, to differentiate between Dendra2 and PAmCherry, peak in the histogram of these proteins is determined (see, Fig. 9.19a). Note that a histogram of the entire set of α-values shows a single peak for each fluorophore. The specimen expressing these two fluorescent proteins show separate peaks is divided amongst the two channels. Accordingly, Dendra2 emission is split such that 48% goes to the reflected light channel and PAmCherry emission is split such that 62% goes through the transmitted channel. This results in a α-ratio peaks at 0.48 and 0.62 for Dendra2 and PAmCherry, respectively [24]. Finally, the merged images (Dendra2 and PAmCherry images) with their respective α-ratio produces the multicolor super-resolved image (see, Fig. 9.19b–d).

9.6 Modern Ultrasensitive Detectors for Single Molecule Imaging

One of the key requirement of SMLM is the ability to detect weak signal (small photon number) from point sources such as single molecules. This calls for detectors that have high SNR, high quantum efficiency, low dark noise , low read-out noise, high detector gain, among others. In this section, we will briefly discuss some of the modern detectors that are routinely used for single molecule detection. These detectors comprises of PMT, CCD, iCCD, cooled CCD, EMCCD, sCMOS, among others. Historically, PMT was first introduced in 1930s, APDs in 1960s, CCD in 1970s, sCMOS in 2000s, EMCCD in 1990s, and SPAD in 2020s. In this section, we will briefly discuss some of these detectors and compare their performance for single molecule detection.

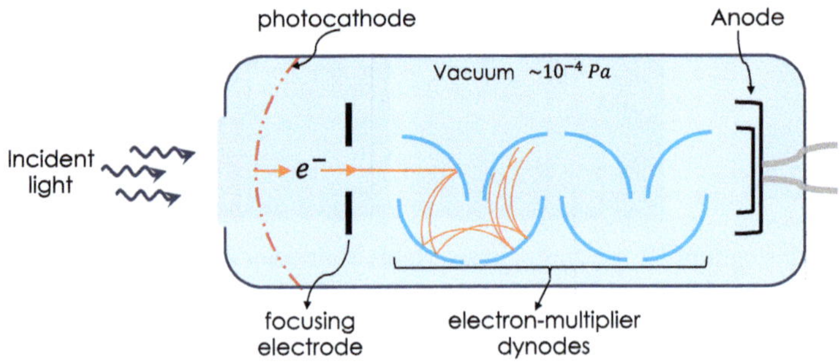

Fig. 9.20 Schematic diagram of a typical photomultiplier tube (PMT)

9.6.1 Photomultiplier Tube (PMT)

PMTs are among the very first sophisticated detectors that came into existence after the discovery of photoelectric effect. PMTs are composed of photo-cathode, multiple dynodes, an anode, all concealed in a vacuum tube at very low pressure.

A typical PMT is shown in Fig. 9.20. Upon incidence of appropriate wavelength photons on photo-cathode, primarily electrons are ejected and accelerated by an electrostatic field towards a series of dynodes. The kinetic energy thus generated by high potential difference between the subsequent dynode (held at more positive voltage potential $\sim$100 eV) knocks off additional electrons (about, 5–10 electrons) at each dynode. This process continues for all the dynodes (approximately, 10–14), ultimately reaching the anode (see, Fig. 9.20), leading to a current gain of about 10^6–10^8. A current in the range of 40 μA is noted for typical GaAs-photocathode based PMT [25]. This current is then fed to the external electronics to further amplify and get it recorded as a single photon. It is the photo-cathode material that determines key properties of PMT such as, quantum efficiency (QE), dark currents, spectral response, among others. Although ultra-sensitive to single photon, PMTs can't discriminate multiple versus single photons. In addition, existing PMTs have better sensitivity in ultraviolet ($QE \sim 0.35$), but poor in visible range (QE < 0.20) [26].

9.6.2 Charge-Coupled Devices (CCD), and Intensified CCD (ICCD)

Charge-coupled device is a huge array of semiconductor substrate coated with photosensitive materials. Each element (pixel) is composed of a metal oxide semiconductor (MOS) capacitor with a surface gate electrode, polysilicon, and charge carrying channel insulated by silicon dioxide as shown in Fig. 9.21. The basic physics behind CCD is the creation of electron-hole pairs upon irradiation by appropriate wavelength photon. This causes an accumulation of charge in the potential well (below the MOS). The movement and release of built-up charge is controlled by an external voltage. Note that the architecture

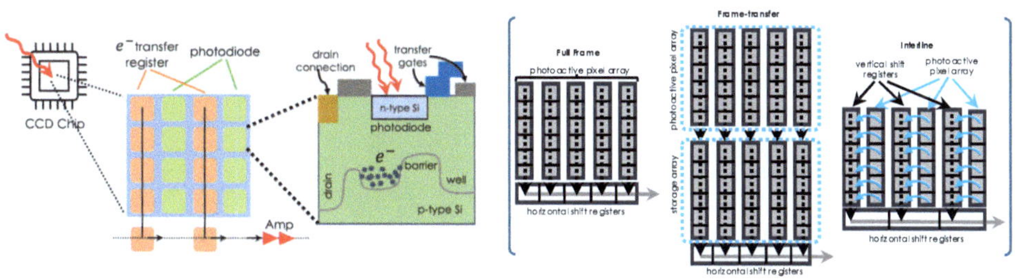

Fig. 9.21 Schematic diagram of a typical charged couple device (CCD) and various architectures (full-frame, frame-transfer and interline transfer)

of electrodes required for charge read-out that act as shift-register for charge transfer. Although many architectures are suggested, there are 3 prevalent types: full-frame, frame-transfer and interline transfer (see, Fig. 9.21).

In full-frame CCD (FF-CCD), the semiconductor region function both as a photoactive element and a charge-transfer device. Upon irradiation of incident photons, the pixel accumulates electric charge during integration. Post integration, the charge packets are moved vertically toward the horizontal shift register. To obtain CCD pixel data, clock signals are timed sequentially to create potential wells and potential barriers in the charge-transferring structure of CCD. One of the major limitation is the need for mechanical shutter that allows the same photoactivable area to store the charge when it is not photoactive.

This is quite different as far as frame-transfer and interline-transfer CCD are concerned. A frame-transfer CCD (FT-CCD) eliminating the need for a shutter by dividing the active area into two equally sized parts, with one part as the normal photoactivable array, and the other is a storage array shielded from light. After integration, charge packets from individual pixels are rapidly transferred to the storage array followed by movements of charge packets to the horizontal shift-registers. This architecture allows FT-CCDs to achieve higher frame rates as compared to FF-CCDs. As far as interline IT-CCDs are concerned, the architecture is able to transfer integrated charge to a storage area so rapidly that smear is reduced to negligible levels. This is accomplished by providing a network of stages adjacent to the photoactive pixels. Post exposure, charge packets in the sensor is simultaneously transferred to the vertical shift-registers. All the three architectures are used in CCDs depending upon the application in hand

9.6.3 Electron Multiplying Charged Coupled Devices (EMCCD)

EMCCD differs from CCD due to its additional feature of on-chip electron multiplying structure. This is positioned between the end of shift-register and external amplification stage (output node) as shown in Fig. 9.22. A typical functional layout of a EMCCD device is shown in Fig. 9.22. The high voltage serial multiplication registers are special due to

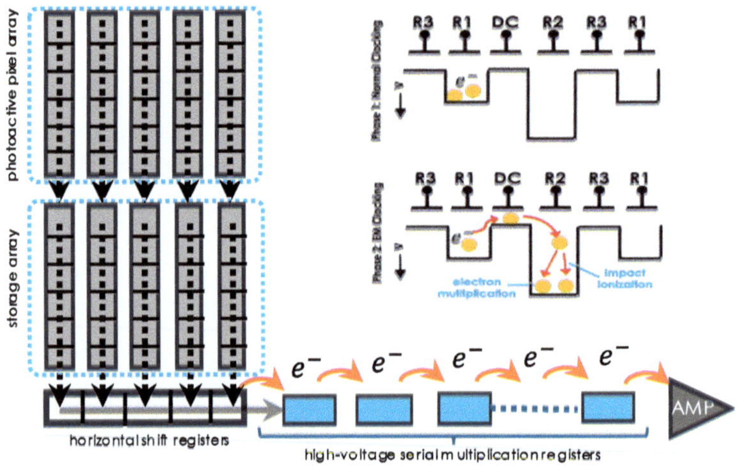

Fig. 9.22 Schematic diagram of a typical electron-multiplying charged couple device (EMCCD) (full-frame), along with normal clocking (phase 1) and electron-multiplying clocking (phase 2)

their ability to increased full-well capacity and acceleration caused to the electrons from one-to-another sequentially by the application of increasing clock voltages at the selected transfer electrodes. When the charge is transferred to the next register by applying a higher voltage, the processes of impact ionization occurs causing the generation of secondary electrons. The gain multiplication register comprises of 4 stages, first three stages are clocked to gain electron multiplication followed by fourth (between phase 1 and 2) which is held at a lower fixed current potential (see inset in Fig. 9.22). As per the schematic, the gates (R1 and R3) are clocked with pulses of potential in the range 5 to 15 volts, with R3 at zero potential. In phase 2, R2 is clocked at much higher potential (in the range of 50 volts), preceded by a gate at low DC-level. The potential difference between the gates cause impact ionization process as electrons are transferred from phase 1 to phase 2. Note that, this is the process that occurs between the conduction and impurity band of a semiconductor, resulting in the production of additional charge carriers. An electron generated by the absorption of a photon gets accelerated due to potential difference before gaining sufficient energy to create electron-hole pair and hence a free electron charge in the conduction band can create another charge. The process continues providing amplification of even a small signal. The whole process causes gain of electrons over a large number of pixels in the multiplication registers, ranging to few hundred to thousands. This is followed by external amplification stage before the actual read-out takes place. Such an on-chip amplification on a solid-state device gives rise to electron-multiplication, resulting in several orders of signal boost.

9.6.4 Scientific CMOS (sCMOS)

Both CMOS and sCMOS cameras/detectors are primarily based on complementary metal-oxide-semiconductor technology (see, Fig. 9.23). Unlike CCD's where all the pixels are fed through one output node, CMOS works in parallel with each pixel having a capacitor and amplifier, and a dedicated analog-to-digitial converter (ADC) for each column. This means electron produced by impinging photon are immediately converted to voltage on the pixel. Subsequently, the voltages from the whole column are send to an ADC. All the ADCs work simultaneously and the entire column is read out, rather than each pixel. This makes CMOS sensors much faster (since they work in parallel) and energy efficient.

Another distinct feature of sCMOS sensor is that they operate in rolling shutter mode. Row-by-row, the image acquisition begins from the top of sensor. This facilitates sensor to acquire images at fast frame-rates with lower read noise. Rolling shutter switch ON and OFF by clearing the charge row-wise thereby allowing for measuring new incident photons. This allows simultaneous acquisition in the current row of a frame while read-out of lower pixel rows of previous frame. Although rolling shutter has a clear advantage, sCMOS do offer global shutter mode for specific application.

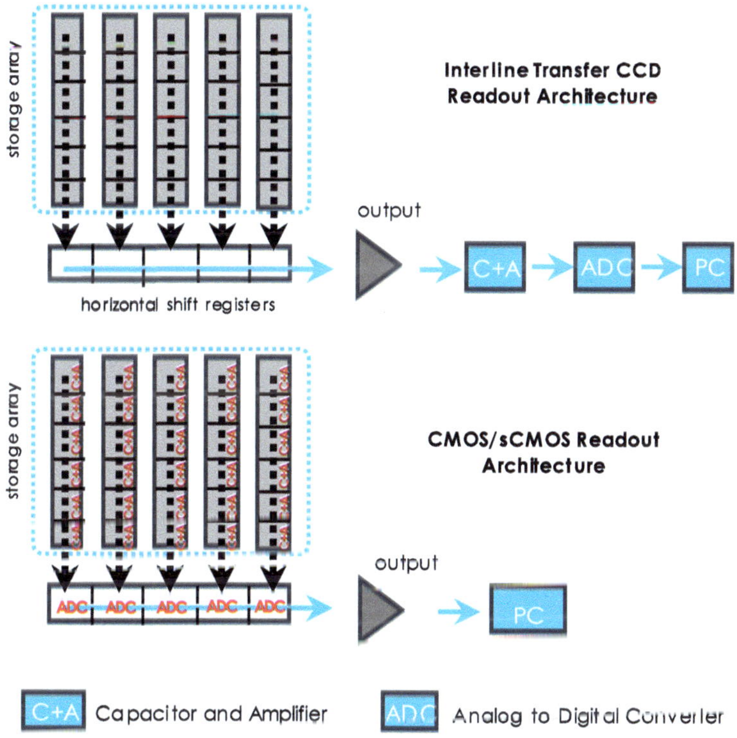

Fig. 9.23 Schematic diagram of a typical CMOS/sCMOS device, along with architecture. For comparison, Interline CCD architecture is also shown

It is important to realize that differences in the architecture of CCD and sCMOS offer both advantages and disadvantages. Here we name a few common ones:

1. High speed: sCMOS are much faster (upto 5 times) than CCDs, which is largely due to ADC associated with each column that processes a fraction of data.
2. Low read noise: The noise associated with sCMOS are 5–6 times less than CCDs (read noise $\sim 6\ e^{-}$). This is largely due to the capacitor and amplifier associate with each pixel.
3. Large field-of-view: In general, fabrication allows large sCMOS sensors, 15–25% larger than CCDs.
4. Power efficiency: sCMOS are known to consume 2 orders less power than CCD, which is largely due to the data acquisition architecture that support parallel acquisition.

Contrary to this, CCDs have some advantages as well:

1. CCDs have less temporal and fixed-pattern noise, which is partially due to more active areas of CCDs.
2. There is a significant rolling shutter artifact in CMOS/sCMOS sensors. In addition, distortion effects are apparent in the acquired images when imaging a moving object that has a timescale of that of the shutter. Moreover, staggered readout between top and bottom row can cause artifacts, and loss of information.

These detectors forms the basis for SMLM enabling detection of weak light emitted by single molecules. With such sophistication and sensitivity these detectors have brought revolution in many fields, with notable contribution in single molecule imaging.

References

1. Abbe E (1873) Beiträge zur Theorie des Mikroskops und der mikroskopischen Wahrnehmung. Archiv für mikroskopische Anatomie 9:413–468
2. Abbe E (1882) The relation of aperture and power in the microscope (continued). J Roy Microsc Soc 2(4):460–473
3. Lord Rayleigh FRS (1879) Investigations in optics, with special reference to the spectroscope. Lond Edinb Dublin Philos Mag J Sci 5th Series 8:261–274. However, in the year 1952, Toraldo di Francia proposed pupil-plane filtering to increase system resolution
4. Toraldo di Francia G (1952) Super-gain antennas and optical resolving power. Nuovo Cim 9:426–435
5. Sheppard CJR, Calvert G, Wheatland M (1998) Focal distribution for superresolving toraldo filters. J Opt Soc Am A 15:849–856
6. Mondal PP, Diaspro A (2008) Lateral resolution improvement in two-photon excitation microscopy by aperture engineering. Opt Commun 281:1855–1859
7. Rogers E et al (2012) A super-oscillatory lens optical microscope for subwavelength imaging. Nat Mater 11:432–435

8. Banerji S, Meem M, Majumder A, Sensale-Rodriguez B, Menon R (2020) Extreme-depth-of-focus imaging with a flat lens. Optica 7:214–217
9. Hell SW, Wichmann J (1994) Breaking the diffraction resolution limit by stimulated emission: stimulated-emission-depletion fluorescence microscopy. Opt Lett 19:780
10. Hell SW, Kroug M (1995) Ground-state-depletion fluorescence microscopy: a concept for breaking the diffraction resolution limit. Appl Phys B 60:495–497
11. Mondal PP (2014) Temporal resolution in fluorescence imaging. Front Mol Biosci 1:11
12. Mondal PP, Hess ST (2017) Total internal reflection fluorescence based multiplane localization microscopy enables super-resolved volume imaging. Appl Phys Lett 110:211102
13. Tokunaga M, Imamoto N, Sakata-Sogawa K (2008) Highly inclined thin illumination enables clear single-molecule imaging in cells. Nat Methods 5:159–161
14. Barak LS, Webb WW (1982) Diffusion of low density lipoprotein-receptor complex on human fibroblasts. J Cell Biol 95(3):846–852
15. Ghosh RN, Webb WW (1994) Automated detection and tracking of individual and clustered cell surface low density lipoprotein receptor molecules. Biophys J 66(5):1301–1318
16. Schmidt T, Schütz GJ, Baumgartner W, Gruber HJ, Schindler H (1996) Imaging of single molecule diffusion. Proc Natl Acad Sci USA 93(7):2926–2929
17. Goulian M, Simon SM (2000) Tracking single proteins within cells. Biophys J 79(4):2188–2198
18. Thompson RE et al (2002) Precise nanometer localization analysis for individual fluorescent probes. Biophys J 82:2775–2783
19. Mortensen, KI, Churchman LS, Spudich JA, Flyvbjerg H (2010) Optimized localization analysis for single-molecule tracking and super-resolution microscopy. Nat. Methods 7:377–381
20. Rieger B, Stallinga S (2014) The lateral and axial localization uncertainty in super-resolution light microscopy. Chem. Phys. Chem. 15:664–670
21. Huang B, Wang W, Bates M, Zhuang X (2008) Three-dimensional super-resolution imaging by stochastic optical reconstruction microscopy. Science 319:810–813
22. Juette MF et al (2008) Three-dimensional sub-100 nm resolution fluorescence microscopy of thick samples. Nat Methods 6:527–529
23. Zamponi N et al (2018) Mitochondrial network complexity emerges from fission/fusion dynamics. Sci Rep 8:363
24. Curthoys NM et al (2013) Simultaneous multicolor imaging of biological structures with fluorescence photoactivation localization microscopy. J Vis Exp 82:50680
25. Photomultiplier tube module H16201-50, Hamamatsu. https://www.hamamatsu.com/jp/en/product/optical-sensors/pmt/pmt-module/current-output-type/H16201-50.html
26. Orlov DA, Glazenborg R, Ortega R, Kernen EJ (2019) UV/visible high-sensitivity MCP-PMT single-photon GHz counting detector for long-range lidar instrumentations. CEAS Space J 11:405

10 Biotechniques and Cell Biology Protocols for Single Molecule Biology

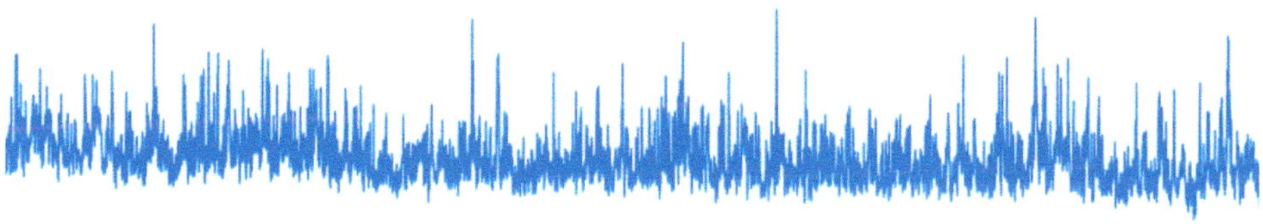

Single molecule biology requires a greater understanding of gene-editing techniques and CRISPR technology, and use of protocols specifically developed for it. This requires the knowledge of gene expression and molecular cloning to construct photoactivable probes (a must for single molecule imaging) by conjugating the gene-of-interest and reporter gene. These recombinant protein based probes enable the study of target biological processes in its most natural setup. This makes use of existing cellular processes (translation, transcription, protein expression and others). The process of molecular cloning is so general enough to cover a large spectrum of biological research (from basic cell to disease biology). In this chapter, we have also incorporated few case studies related to Influenza (type B) and Dengue (type 2). Rest of the chapter is dedicated to a more general framework of single molecule protocols essential for cell and disease biology. Specifically, details related to the construction of recombinant plasmids and cell transfection protocols are given. Protocols associated with cell culture, fixing, and imaging are also incorporated. In addition, preparation of lab microscopes for single molecule imaging and a list of essential optical components are outlined. We also discussed the imaging modes (Widefield/TIRF/HILO), its calibration, data acquisition and image analysis, and the details of sensitive detectors (EMCCD/ sCMOS).

In this chapter, we will discuss methods to monitor gene expression with single molecule sensitivity in live cells. The method ensures detection of single fluorescent protein in the presence of auto-fluorescence background and other sources of noise (such

P. P. Mondal, S. Hess, *Single-Molecule Super-Resolution Optical Microscopy*,
https://doi.org/10.1007/978-3-032-13442-4_10

as, electronic and thermal noise). This is essential for a number of studies and aided by the recent advancement in gene expression involving photoactivable proteins. Over the years, dedicated cell biology protocols and biological techniques are developed to study biology at single molecule level. Here we will discuss some of the widely adopted biotechniques and protocols necessary for single molecule based super-resolution imaging.

Single molecule imaging is multidisciplinary in nature that requires a wide range of diverse protocols (gene expression, preparing the microscope, detection modes, detector calibration and data analysis) for qualitative imaging. Among these, photoactivable protein probes play a significant role. In general, the expressing protein-of-interest and photoactivable protein expressing gene are combined using molecular cloning techniques. The recombinant plasmid is internalized in a cellular system using transfection method where it get expressed. Subsequently, the cell machinery processes the plasmid DNA to generate the targeted protein-of-interest. The cell is then imaged by an optical system and the single molecule data is then computationally processed and analyzed. The analysis and computational techniques are of critical interest in determining the final super-resolved image and to carry out single molecule analysis. These analysis forms the basis for studying fundamental processes in cell biology and understand disease biology.

In this chapter, we will elaborate on the biological techniques that enable sample preparation for single molecule imaging. In addition, we will discuss proven cell biology protocols, image processing methods and data analysis techniques. It is expected that, this chapter will enable reader to prepare their own biological samples and carry out single molecule imaging.

10.1 Basics of Molecular Cloning Gene Expression

Molecular cloning is a process of constructing recombinant DNA, and its insertion in a vector (plasmid) for propagating within a host cell, like bacteria, mouse cell, human cell etc. The machinery within a cell is capable of expressing these recombinant DNA into protein of interest, encompassing many known processes from endocytosis to translation of a specific gene or fluorescent protein. The process of molecular cloning can be used to prepare and express a protein of interest and its function in a cell.

The basic cloning method involves the following important steps:

1. The isolation of gene of interest (target DNA) and its amplification using PCR or restriction-digestion from parent vector to obtain enough material for cloning.
2. The DNA is purified and restriction enzymes are used to cut at specific locations. The same restriction enzyme is used to cut the vector DNA to generate compatible ends.
3. Then, both the vector and target DNA are mixed in appropriate ratio in the presence of DNA ligase. Note that, the DNA ligase stitch the target DNA to the vector backbone.

4. The resultant recombinant plasmid is then transformed into the host system. Often bacteria is used as a host system to grow large quantities of plasmid containing the cloned DNA.
5. PCR using specific primers, restriction fragment patterning, and DNA sequencing can then be used to validate the recombinant nature of the cloned material.

We will understand individual processes in detail in the subsequent sub-sections.

10.1.1 Cloning an Organelle Reporter Construct

Cloning a gene from an organism and express it in a cell (prokaryotic or eukaryotic) requires isolation, purification, amplification, analysis and sequencing. Here are the key steps:

Cloning This is the process of creating a genetically identical copies/replica of short stretches of DNA. This technique is also known as molecular clonning. In short, the technique allows for the generation of multiple copies of DNA using the biological machinery of a bacterial cell by inserting it (DNA fragment) into a plasmid vector. It must be ensured that, the DNA fragment is in a form that can be expressed/copied.

Cloning and Expression Vector The cloning and expression vectors are two types of vectors used to carry target DNA into the cell. Cloning vector is used to transport foreign DNA into the cell and its multiplication inside the cell, whereas expression vector contain appropriate regulatory sequence to ensure maximum gene expression. Cloning vectors can be plasmids, bacteriophages or cosmids, however the expression vectors are plasmids only. Here we discuss only expression vectors as we are dealing with recombinant protein expression (reporter gene).

Ligation For molecular ligation, the resultant DNA fragments contains blunt ends or overhangs that allows joining of vector DNA with compatible ends. This process is called *molecular ligation*.

Cloning Site Vectors have sites that facilitate a gene to be inserted or replaced. In general, there may be multiple cloning sites also known as polylinker. First, these restriction sites are cleaved by restriction enzymes which is then followed by ligation where a PCR-amplified target gene (also digested with the same enzyme to generate sticky ends) is ligated to the vector backbone using DNA ligase.

Restriction Enzymes As the name suggests, these enzymes are restricted to cut or digest DNA. They are able to recognize specific DNA sequences and can cut them at specific

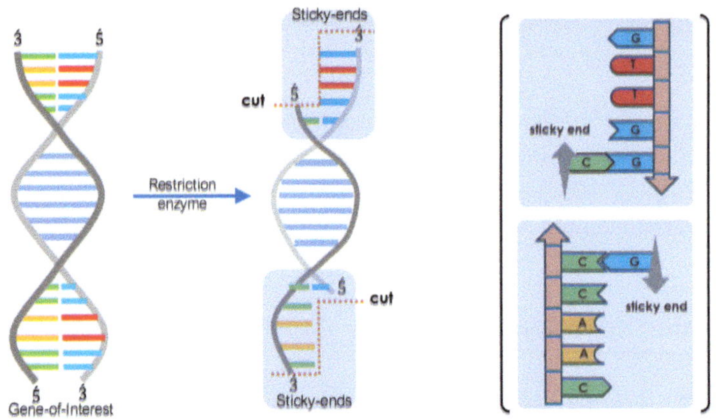

Fig. 10.1 Sub-cloning of a typical DNA reporter construct: Generation of sticky ends at the ends of Gene-of-interest by restriction enzyme, allowing it to stick on vector backbone

site. In general, they cut DNA sequence into a staggered pattern where the overhangs are complimentary as shown in Fig. 10.1.

Selectable Marker A selectable marker in the vector allows the selection of positively transformed cells. Specifically, antibiotic resistance such as kanamysin and amplicillin is often used as a marker which gives resistance to the group, thereby enhancing selectivity and growth of transformed cells in a minimal growth medium.

Reporter Genes These genes are used in expression vector to screen and identify successful clones using specialized features. These features facilitate and confirm the production of recombinant proteins. For example, one of the well-known fusion gene is green fluorescent protein (GFP) that is widely used for screening. Figure 10.2 shows a recombinant plasmid with a GFP reporter gene.

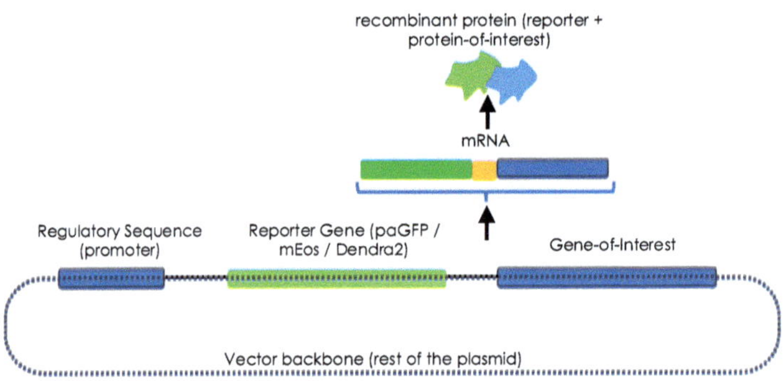

Fig. 10.2 Schematic of a complete cloning process with key components

Promoter Promoters are the elements that are responsible for the expression of the cloned target gene. For a successful expression of a targeted gene, the expression vector needs to contain a promoter and ribosomal binding site (RBS). Ideally, the target DNA is inserted into a site which is under the control of a promoter. This helps the expression of the target gene in the chosen host (E. coli/mammalian cell/yeast). The function of the promoter is to ensure controlled production of proteins as per requirement.

Finally, choosing the right cloning vector is essential for stable maintenance of DNA inserts. This depends on a number of factors, such as, insert size, copy number and cloning method. Of many, plasmids which are autonomously replicating circular DNA are among the most commonly used cloning vectors. These plasmids often have large size of ~15 kb, and high copy number (~500–700 copies per cell) that help produce a high yield of recombinant plasmid. This is beneficial since it improves SNR thereby enabling single-molecule super-resolution fluorescence microscopy.

Figure 10.2 summarizes the entire process of DNA cloning in a step-by-step manner with critical components.

10.2 Selection of Fluorescent Protein (FP) for Super-resolution

With the discovery of green fluorescent protein and its derivatives, it made possible to use it as gene expression reporter [1]. Here are the important steps in the FP selection process:

10.2.1 Construction of Recombinant Vector Expressing Gene-of-Interest/Reporter Plasmid

The first step towards extended live cell imaging with a resolution of single molecule is to construct the reporter plasmid. This requires the conjugation of targeted DNA (gene-of-interest) to the photoactivable gene, under the control of highly-repressed CMV promoter. This ensures high copy number of the recombinant protein. Once expressed, the targeted protein quickly fold, mature and translocate into the respective organelle. These proteins are then detected individually to build-up the single molecule map of the respective organelle or associated process (such as, clustering and migration). A caution of note is that these proteins are expressed at such a low-level that they are undetectable by traditional methods (western blotting, mass spectrosmetry etc). In the subsequent sections, we will discuss these processes in details.

10.2.2 Selection of Photoactivable Fluorescent Protein

Traditionally, GFP and its derivatives are used as a choice of interest for gene expression and protein localization. For real-time single molecule imaging of gene expression the following factors are critical:

1. The photoactivable fluorescent protein matures fast so as to enable observation of fast dynamics of gene expression activity.
2. The fluorescent protein must be bright enough so that it can be detected over the autofluorescence background of the cell. This will allow the observation of single molecules.
3. The emission of fluorescent probe is red-shifted so that it can be easily distinguished from cell's autofluorescence.
4. The fluorophore has moderate to high photostability. This ensures that photo-induced degradation (photobleaching) occurs but only after it is observed to emit many photons, thereby allowing single observation of individual molecule.

10.2.3 Selection of Probe Sequence

The selection of fluorescent probe is critical. A localization probe sequence is the one that confines the fluorophore to the designated target. For example, a membrane localization probe sequence helps targeting to membrane. In general, a fluorophore molecule will freely diffuse, resulting in the spread of fluorescence. Therefore, the localization sequence is fused to the fluorophore gene to ensure confinement. So, when a reporter protein molecule is expressed, it is to the target location. This enhances its signal-to-background ratio, allowing the detection of single reporter molecule, and brings in selectivity.

10.2.4 Gene Engineering and CRISPR Technology

A large part of gene engineering uses recombinant DNA technology to modify organism's DNA to obtain desirable function. Molecular cloning is the most common method used for the addition of foreign recombinant DNA vectors as discussed in the previous section. An organism receives recombinant DNA is called *genetically modified*, whereas the induction of foreign DNA derived from a foreign species in a host organism is called *transgenic*.

CRISPR (clustered regularly interspaced short palindromic repeats) is a new technology and represents DNA sequences found within the genomics of organisms (such as, bacteria). Special enzymes such as, $Cas9$ uses CRISPR sequences as a guide to recognize specific stands of DNA and cleave the same. A new technology called $CRISPR - Cas9$ that employs $Cas9$ enzyme in combination with CRISPR sequences is used to edit genes within organisms. This editing process has shown great potential in medical and clinical research for the treatment of diseases. Further, details about this emerging technology can be found in Refs. [2, 3].

10.3 Organelle Reporter Construct for Cellular Level Study

In this section, we will discuss the subcloning protocols specifically suited for cell based study for applications in disease biology. Although the protocols largely remain same, but we will be taking specific example of Dengue (type 2) and Influenza (type 2) as a case study.

10.3.1 Case Study: Recombinant Plasmids for Dengue (Type 2)

As per literature and studies carried out over the decades, the non-structured protein complexes that is preserved throughout the Dengue viral cycle are $NS2B$ and $NS2B3$ (a complex composed of NS2B and NS3 protein) [4, 5].

Hence, taking NS2B3 as the protein-of-interest, we will develop strategies to create a photoactivable construct. The process begins with the selection of target protein-of-interest and photoactivable probe in the desired spectral range. The selection of probe excitation sources (activation and excitation lasers) with wavelengths associated with the activation and excitation of the probe. As a case study, we take NS2B3 as the protein-of-interest and PAGFP/mEos as the photoactivable protein probe. The next step is the conjugation of these proteins (also called recombinant protein) using molecular cloning technique as discussed in the previous section.

10.3.2 Protocol for the Creation of Recombinant Plasmid and Confirmation

We take PAGFP/mEos as the photoactivable protein probe and wish to integrate with NS2B3 protein complex of the Dengue virus. The operational details of the light sources and optics necessary for this probe (its activation and excitation) will be discussed in the next section.

The subclonning protocol for the above conjugation process is as follows:

Step 1: Obtain the base plasmid containing gene of interest either from a collaborator or from international repositories such as AddGene [6]. In our case we obtained the base plasmid $pcDNA - DENV2 - NS2B3$ from Addgene which is a Gift from Alan Rothman [7].

Step 2: Obtain a suitable photoactivable protein i.e, PAGFP for activation and excitation at 405 nm and 488 nm, respectively or mEos for activation and excitation at 405 nm and 561 nm, respectively. In our case, we decided to work with both PAGFP and mEos.

$PAGFP-C1$ plasmid which is a gift from Jennifer Lippincott-Schwartz [8] is chosen as vector back bone. The mEos fragment is from mEos3.2-TOMM20-N-10, a gift from Michael Davidson [9].

Step 3: Subclonning of NS2B3 (insert) at Pst1- BamH1 restriction site of PAGFP-C1 plasmid (vector backbone) is carried out. Accordingly, NS2B3 sequence with PstI-BamHI overhangs were generated by PCR using NS2B3-specific oligonucleotide primer, NSPstFP, with PstI site as the forward primer and NSBamRP with Bam-H1 site as the reverse primer. PCR was performed using Phusion™High-Fidelity DNA Polymerase (Thermo Fisher Scientific, India). PAGFP-C1 plasmid was also digested with similar set of restriction enzymes to generate compatible ends. Both vector band and insert band were eluted after running on 0.8% agarose gel using Genejet gel extraction kit (Thermo Fisher Scientific, India).

Step 4: The gel eluted products (Both insert and vector backbone) were ligated for 1 h at 22 degree using enzyme ligase (Thermo Fisher Scientific, India) at 1:3 vector to insert ratio.

Step 5: Transformation were performed using the standard protocol (Sambrook and Russel 2001) in E.Coli strain DH5α.The recombinant colonies were master plated and isolated the plasmids.

Step 6: The presence of NS2B3 fragment in recombinant plasmid PAGFP-NS2B3 was confirmed by polymerase chain reaction (PCR) and restriction digestion analysis. The PCR of PAGFP-NS2B3 using NS2B3-specific primers gave an amplicon around 2300 bp and restriction analysis using the restriction enzyme pair Pst 1 and Bam H1 released a fragment of 2300 bp along with 1Kb ladder. The recombinant nature of PAGFP-NS2B3 was further confirmed by sequencing.

Step 7: For the plasmid, the $mEos$ with $NheI-PstI$ overhangs were generated by PCR using oligonucleotide primers, $mEosNheFP$ with $NheI$ site as forward primer, and $mEosPstRP$ with $PstI$ site as reverse primer. The amplified product after restriction digestion with NheI and PstI was ligated into PAGFP-NS2B3 plasmid backbone by removing PAGFP at NheI-PstI site followed by transformation. The list of the sequence of primers used can be found in Ref. [5].

Step 8: Similarly, the presence of NS2B3 fragment in recombinant $mEos - NS2B3$ was confirmed by PCR restriction digestion analysis and sequencing. Sequence analysis confirmed the intactness of the reading frame while fusing NS2B3 with the fluorophore (PAGFP/mEos).

Step 9: Results from PCR, restriction analysis and sequencing along with recombinant vector map is shown in Figs. 10.3, and 10.4.

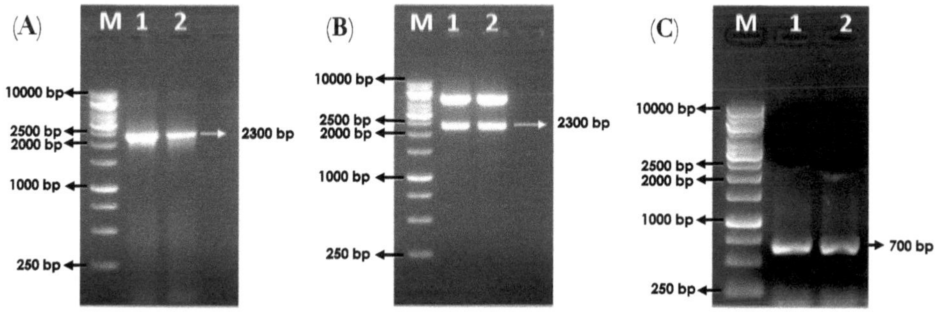

Fig. 10.3 PCR amplification of NS2B3 protease

10.3.3 Case Study: Recombinant Plasmids for Influenza (Type A)

In this case study, the reason for choosing a specific gene (Hemagglutinin, or HA) is due to its conserved nature throughout the Influenza viral cycle. Here, HA is taken as the protein-of-interest, and $Dendra2$ as the reporter photoactivable protein. The corresponding recombinant protein is obtained by conjugation following protocols discussed in the previous section.

10.3.4 Protocol for the Creation of Recombinant Plasmid and Confirmation

A similar protocol to that of Dengue (type 2) plasmid is followed for constructing recombinant plasmid for Influenza (type A). However, it may be noted that Influenza plasmid is the first to be generated i.e, before the Dengue (type 2) [10].

pDendra2-HA and pPAmCherry-b-actin, also known as Dendra2-HA and PAmCherry-actin, respectively are based on the pEGFP-N1 vector (Clontech, Mountain View, CA). The coding DNA of the hemagglutinin gene of the A/X-31B (Puerto Rico/8/1934-Aichi/2/1968) (H3N2)strain of influenza virus was used in the pDendra2-HA plasmid. pPAmCherry-cofilin (also known as PAmCherry-cofilin) was made by cloning human (non-muscle) cofilin 1 between BamHI-AgeI sites of the pEGFP-pPAmCherry vector. This vector is the same as that used for pPAmCherry-b-actin, with pEGFP replaced by pPAmCherry in pEGFP-N1 (Clontech). The step-wise protocol is similar to that outlined in the previous section.

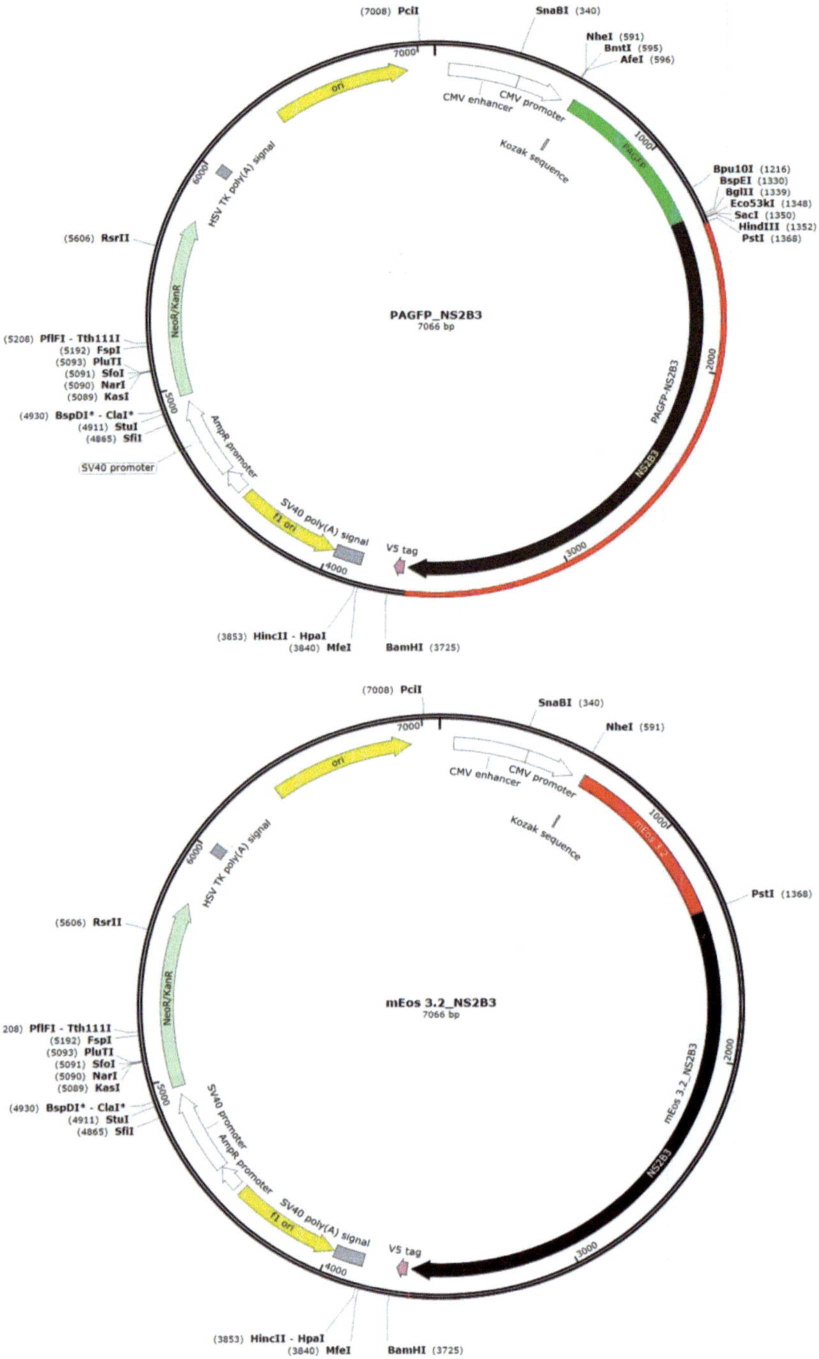

Fig. 10.4 (**a**) Recombinant plasmid map of PAGFP-NS2B3 with PAGFP (marked as green) and NS2B3 (marked as black). (**b**) Recombinant plasmid map of mEos-NS2B3 with mEos (marked as red) and NS2B3 (marked as black)

10.3.5 Protocol for Transfection of Cell lines with Recombinant Plasmid Containing Gene of Interest

In general, the transfection is a process by which foreign nucleic acid is introduced into eukaryotic cells to alter its genetic makeup. So the technique has wide spread applications in studying cellular processes and disease mechanisms. The selection of a of a suitable/standard cell line with a specific recombinant plasmid depends on the application. For our case, the specific process involve, the transfection of NIH3T3 cells (mouse embryonic fibroblast cell line) with recombinant dengue plasmids ($paGFP - NS2B3$ and $Dendra2 - NS2B3$) to study the interaction of dengue NS2B3 protease with the cellular organelles. The steps involve are as follow:

Step 1: The cells are thawn from LN2 storage container and centrifuged with culture media of equal amount to pellet down the cells.The supernatant is removed and cell pellet is resuspended in complete growth media containing DMEM (Dulbecco's Modified Eagle Medium) supplemented with 1% antibiotic Penicillin-Streptomycin and 10% FBS (all cell culture reagents purchased from Gibco, USA).

Step 2: The resuspended cells are cultured into T25 flask with 3ml of culture media and incubated at + 37 degree Celsius with 5% CO_2. The cells are then splitted after 2 days or when the confluency is 80% whichever is early.

Step 3: The cells were cultured (in CO_2 Incubator) for two consecutive passages to ensure healthy growth.

Step 4: During the third passage, the cells ($\sim 10^5$) are seeded in 35 mm dishes with coverslips (No. 0) (Blue star, India) and then incubated for 12–14 hours to allow them to reach 60–70% confluency. The cell medium is removed and washed two times with 1 ml of 1X PBS to remove cell debris.

Step 5: Transfection was performed using 1μ g of plasmid DNA with Lipofectamine 3000 (Invitrogen, USA) according to manufactures instructions.

Step 6: The cells are incubated with the complex and sufficient media to keep them hydrated for 24 h. The culture dish is loosely covered by aluminium foil to avoid unnecessary light exposure.Subsequently, the entire media containing the lipofectamine-DNA complex is removed, and fresh culture media is added. Then the cells are incubated for another 24 h to have sufficient expression of recombinant proteins.

Step 7: After 24 h of transfection the cells are fixed, the medium is removed, and washed 2 times by 1X PBS. Subsequently, 3.7% Paraformaldehyde was added to the dish and kept for 15 min in the biosafety hood.

Step 8: Finally, paraformaldehyde solution is pipetted out and washed 2 times with 1X PBS. The mounting media Fluorosave (Invitrogen, USA) is added to the cleaned glass slide and coverslip is placed upside down on top of thc mounting media. The glass slide is left for 24 h for drying.

Step 9: The fixed cells are made airproof by applying nail paint on the edges to avoid drying-up, keeping up the moisture content and for long time preservation. The glass slide is wrapped in aluminum foil and preserved at 4 °C.

Step 10: Note that all the process is carried out in a Biosafety hood and CO_2 incubator. The cell is cultured in a 35 mm cell culture disc (with No. 0 coverslip at the bottom).

10.4 Cell Based Targeted Study and Protocols

10.4.1 Cell Culture and Maintenance

To begin with, the cells were thawn from preserved specimen in liquid Nitrogen container LN2 dewar (− 196 °C). The cells are maintained in Dulbecco's Modified Eagle Medium (DMEM) complete media (89% of DMEM medium (Gibco, Thermofisher, USA), 10% Fetal bovine serum (FBS) and 1% Penecilin-Streptomycin (Gibco, Therofisher, USA)) at 37 °C and 5% CO_2 (CO_2-incubator, Thermo Scientific, USA). For healthy growth, the cells are seeded at 10^5/mL concentration on a cover slip(No. 0, thickness 0.1 mm) (Bluestar, India) in a 35 mm petridish (Biofil, India) or on a live imaging dish (MatTek, USA) and were cultured for 12–48 hrs, depending on the study.

10.4.2 Cell Transfection and Fixation Protocol

The recombinant plasmid is used to transfect cells (say, NIH3T3 cells—mouse embryonic fibroblast cell line). The transfection is performed using plasmid DNA with Lipofectamine 3000 (Invitrogen, USA) according to manufacturer's protocol. The cells can be cultured on a coverslip (No.0) (Blue star, India) in a 35mm dish with a density of 10^5 cells/ml. After 12 h of incubation, the cells are transfected with the respective plasmid. In general, the tranfection requires 12–48 hrs for the protein to get expressed. This can be easily seen in an inverted fluorescence microscope (with blue light (470–490 nm) illumination for Dendra2 and right set of filters). Post-transfection, cells are washed thrice with 1X PBS, and cells are incubated and fixed with (W/V) paraformaldehyde for 15 min. This is followed by washing with 1X PBS two times. The cover slip is removed from the 35mm dish and placed on a clean glass-slide with mounting media fluorosave (Thermofisher, USA). The fixed cells are made airproof by applying nail paint on the edges for long-term preservation and further investigation.

10.5 Live Cell Imaging Protocol

Super-resolution imaging of live cells can be performed in a straight forward manner. Preparation of the microscope system, automation and related protocols can help in obtaining high quality image acquisition, specifically needed for capturing dynamic cellular events with high temporal resolution. It is often observed that increasing SNR requires longer exposure at low frame-rate acquisition, but causes phototoxicity, thereby a fine balance is often necessary to accurately record dynamic events in live cells. Below, we highlight few equipments that can be optimized during imaging.

10.6 Preparing the Microscope

1. **Vibration Isolated Table:** An optical table with isolators that are capable of isolating vibrational noise is a must. These kind of isolating platforms are essential for detecting and tracking single particles/molecules for single molecule experiments.
2. **Semi-motorized inverted fluorescence microscope**: For experimentation, it is often advised to procure research microscopes with multiple ports that can be used for light and other device coupling. It may beneficial to have microscopes with auto-focus, precision stage positioning (typically, 0.5 μm step-size), shutter, fluorescence filter cube. With this basic structure of the microscope, other automation may be needed for multicolor and time-lapse imaging.
3. **Light sources/lasers**: In general, different light sources are used for different purposes. Broadband light sources such as, Xenon arc lamps/LED lights are used for transmission imaging and placing the sample for initial inspection. However, lasers with specific wavelengths are preferred for activation/ excitation of the specimen, and along with specific filters provide a complete system for SMLM imaging. Multiple lasers are often preferred for simultaneously imaging of multiple organelles/processes in a live specimen.
4. **Sensitive Detector:** One of the defining components necessary for single molecule imaging is the detector that is sensitive to even single photon. Cameras such as, EMCCD, cooled CCD, sCMOS and SPAD are the suitable ones. These detectors have demonstrated low read-out noise, high quantum efficiency, fast frame-rates, low thermal noise among others that are of utmost importance for single molecule detection.
5. **High NA Objective Lens:** Objective lens with high NA (typically > 1.4) are beneficial for collecting more light which ultimately improves the image quality and resolution in single molecule imaging. These high NA objectives are multifunctional in the sense they can be used in all three modes of imaging i.e, Widefield, TIRF and HILO. In addition, these sophisticated objectives comes with correction for coverglass thickness and chromatic aberration.
6. **Live Cell imaging disc:** Most of the live cell imaging is performed in a closed 25/35 mm disc with the bottom attached to a coverslip. This is helpful for its compatibility

with the objective lens that has small working distance. In addition, these close discs help maintain the cell medium at 37 °C for live cell imaging. However, an on-stage incubator is preferred for carrying out long-term imaging.

7. **Filters, BS, and other accessories**: Separation of the signal of interest (i.e, fluorescence not autofluorescence or background) form the background is a crucial step that is performed by filters, beam-splitters. Especially, the beam splitters (embedded in the filter wheel/case of modern microscope) are the one whose primary job is to distinguish fluorescence from the excitation light sources (405 nm and 561 nm for super-resolution imaging). Another kind of filters that completely eliminates the Gaussian background of laser light sources are notch-filters (placed just before the camera). Apart from these, longpass filters and bandpass filters may be used for better filtering of signal (fluorescence).
8. **Image Analysis and Softwares**: The code/software used for data collection must be compatible with microscope, camera, shutters, optically-tunable lens and other devices. Automation is a must when considering high quality imaging. This calls for dedicated computers and its coupling with devices via Arduino or Labview or other softwares as preferable. MATLAB is a key software platform for self-developed codes for data processing. The microprocessor must be fast enough to run demanding math functions (FFT, spot detection matrix-multiplication etc, deconvolution and tracking). Finally rendering softwares such as, ImageJ and FIJI are excellent data analysis tools.
9. **Other Useful Devices**: Live cell imaging demands additional devices and calls for complete automation of the system for prolonged super-resolution imaging. These are autofocus mechanism, automatic sample stage and digital micromirrors. Specifically, the autofocus mechanism is essential which is a continuous reflective interface close-loop feedback system for zero-drift compensation. This is a key element for obtaining drift-free high quality images for long duration/time-lapse imaging. Automatic sample stage gives the flexibility for volume imaging of specimen that needs repeated visit to the initial x, y, z positions. In addition, this has become a necessity for image-stitching applications. Finally, digital micro-mirrors are important for deflecting or modulating the excitation beam for applications requiring change of incident beam or intensity.

10.6.1 Microscope Configurations (Widefield/TIRF/HILO)

The microscopy system needs calibration for it to be efficiently used in Widefield or TIRF or HILO and for application requiring toggling from one to another mode. One example is super-resolution microscopy where certain applications require it to efficiently toggle between Widefield and TIRF mode. In the next section, we will discuss calibration and alignment of a home-built super-resolution microscope system.

10.6.2 TIRF System Calibration and Alignment

With the assumption that the user is familiar with the basic operation of an epifluorescence microscope, we set on to lay down the protocols for TIRF microscopy:

Step 1. Switch on the inverted semi-automated fluorescence microscope, the EMCCD camera, and the excitation lasers (externally coupled to microscope).

Step 2. Bring the specimen (fluorescent dyes/beads dispersed on a coverslip) into focus under transmitted light (placed on the top of most inverted microscopes).

Step 3. First turn off the transmission light and allow the laser light to fall on the specimen (either on a coverslip or live cell imaging disc with coverslip at the bottom). Observe the specimen in an epifluorescence illumination mode by adjusting the mirror-lens combination (placed on a linear translator) at an incident angle, $\theta = 0$ degrees.

Step 4. Now, move the translator so that the illumination beam moves to the periphery of the objective lens (of high NA i.e, $NA > 1.42$). Continue this till the laser beam reflects back along the same path i.e, the beam undergoes totally reflected at the interface of glass coverslip ($RI \sim 1.32$) and immersion oil ($RI \sim 1.55$). Observe the image by eye with laser held at high intensity. Under this condition, one must observe fluorescent spots. Ensure that all the spots are uniform size and intensity.

Step 5. One can change the light path at the back-port of the microscope slightly. Accordingly, adjust the power of the laser, the sensitivity of the EMCCD camera. The sensitivity (EM gain) of the camera must be gradually increased by looking at the image on the monitor. In addition, increase the intensity of the laser gradually.

Step 6. Now, place a new sample (say, fixed cell) and bring it into the field. Adjust all the parameters a bit more for the new sample since fluorescence proteins are weaker than dyes. If required go back to *Step 4–5.*

Step 7. At this stage, the system is ready to take image in TIRF mode. Start acquiring the images from the camera software and process them using home-grown scripts. Self-written codes are better as this allows incorporation of functions that may be specific to the study under investigation.

10.7 Single Molecule Data Acquisition, Analysis and Image Processing

In general, the data acquisition is performed by the camera software. The EMCCD camera requires the user to choose the parameters in the following order:

(i) **Cooling:** The EMCCD camera is equipped with the chip-cooling hardware (Peltier cooling system). The water cooler must be switched on so that the chip temperature can be brought down. Alternately, the camera can be operated in air-cooling mode which requires air-conditioned room locked to $<18\,°C$.

(ii) **Frequency:** Modern EMCCD cameras comes with data acquisition rate (1–10 MHz). A proper selection of this is essential from quality data collection. Too fast may reduce the number of photons and too slow may be insufficient for observing temporal changes in single molecule blinking. For most of the super-resolution imaging, 1 MHz seems to be a good choice.

(iii) **Exposure Time:** The detectors allows exposure from as low as few hundred microseconds to as high as few tens of second (large dynamic range). The choice of exposure time depends on the blinking period of the target photoactivable molecule. For example, Dendra2 and PamCherry has an approximate average blinking period of 30 ms.

(iv) **EM Gain:** This factor determines the gain achieved by the onboard electron multiplying system. The minimum is 1 (no gain), whereas maximum can be upto 1000. This means the optical signal (number of photons) can be amplified upto 1000 times before being detected by the external circuit. This gain is key to record the feeble signal (300–1200 photons) of single molecules in the presence of background (200–500 photons). However, note that noise also gets amplified and so a balanced gain must be used. In practice, a gain of $\sim$280 may be optimal for most of the single molecule imaging experiments.

(v) **Other parameters:** Super-resolution microscopy requires the setting of other parameters such as, the active recording area (pixels) and the number of frames. In general, one can use the full frame (system default) or choose the area of interest (by reducing the window size). For visualization purpose, one can use live histogram to analyze the data.

(vi) **Other parameters:** Finally it is important to save the camera settings configuration for future use.

10.7.1 Data Processing Algorithm

The algorithm used for the identification of single molecules and rendering the super-resolution map has the following major steps:

Step 1. Background subtraction: Once the data is loaded to the PC/server, the first step is to reduce the background. This is achieved by selecting a window just outside the region of interest and computing the average background noise. This value is then subtracted from all the recorded frames. Rolling ball or temporal median filtering are also good options.

Step 2. Spot detection: The noise-subtracted frames are then analyzed serially for the presence of bright spot. This is implemented by comparing the blob with a diffraction limited spot ($\sim\lambda/2$) or more considering additional magnification in the detection module). If the blob is nearly equal to the spot, then it is considered as the signature of a single molecule. Note that, smaller spots may represent noise and larger spot may correspond to overlapping single molecule PSFs. Hence all such ambiguous spots are discarded. Finally, all the acceptable blobs are identified and sent for further processing.

Step 3. **2D Gaussian fit:** The selected blobs are then subjected to 2D Gaussian or actual PSF fit (available in MATLAB as a inbuilt function) with all the default parameters. The corresponding fit-parameters are also stored for quality assessment.

Step 4. **Centroid determination and localization precision:** From the fitted 2D Gaussian, pixel-to-photon conversion is carried out. This depends on the expansion factor and gray-level histogram plot. Finally, the total number of emitted photons per single molecule PSF is determined and its centroid is identified. From the number of photons (N), the localization precision ($l_p = r_0/\sqrt{N}$) is calculated for all the recorded molecules. Here, terms related to background and pixel size are ignored.

Step 5. **Data Rendering and Visualization:** The final step is rendering the data (centroid and l_p of single molecules) to construct the super-resolved map on a new image grid with an expansion factor of 15 (say). Accordingly, the centroid is fixed for the particle on the new expanded grid. A new Gaussian is generated with its center at the estimated centroid location and standard deviation as l_p to represent the molecule. In addition, the region-of-interest (defined by the pixel values both along X and Y) is noted. In some cases, it is more informative to visualize the map with an upper size limit (say, 20 nm), where molecules larger than 20 nm is represented as 20 nm-sized Gaussian.

10.7.2 Data Analysis for Biophysical Parameter Estimation

Once the single molecule blob is recognized and the basic parameters (centroid and its l_p) are determined, the data can be subjected to statistical analysis to determine its distribution (say, in a cell). This will give us important information related to its density, area of distribution, and other parameters. In this section we will discuss some of these statistical techniques that are used for cluster analysis.

Point Clustering This method uses Euclidean distance, $J = \sum_{i,j} \sqrt{(x_i - x_j)^2 + (y_i - y_j)^2}$ to identify the next point in the cluster. Here points are refereed to the positions of single molecules in the super-resolved map. Depending on the distance cut-off, its is determined if the next point belongs to a cluster. As an example, a cut-off of 30 nm and a minimum number of 200 points are generally chosen for the ensemble to be called as a valid cluster. This technique has the advantage of accounting only for the clustered molecules/points.

Gaussian Mixture This method requires the identification of a cluster of molecules by fitting a Gaussian function , $G(\mu_i, \sigma_i) = \frac{1}{\sqrt{2\pi}\sigma_i} \exp\{-(x-\mu_i)^2/2\sigma_i^2\}$, where, μ_i and σ_i are respectively the mean and standard deviation of the Gaussian function. In addition, the method uses iterative expectation maximization algorithm to converge to the most likely solution.

K-means In this method, the mean of the cluster is placed at a random location and each point is assigned to the nearest mean, primarily based on the least-squared Euclidean distance, $J = \sum_{j=1}^{k} \sum_{i=1}^{n} ||p_i - c_i||^2$, where $||p_i - c_i||$ is the Euclidean distance between the point p_i and the mean c_i. This is an iterative process where each centroid ($c_i = \frac{1}{N_i} \sum_{j=1}^{N_i} x_j$) is recalculated until convergence. In practice, this can be practically implemented using the built-in k-means $++$ Matlab function.

DBSCAN This is a density-based clustering method designed for identification of clusters. To begin with, the neighborhood distance ϵ and a minimum number of points m_p is specified. For a point, a set of points within the neighborhood distance of ϵ are determined based on Euclidean distance. If the number of neighbors are less than m_p, then the current point is labelled as noise, else it is considered as core point belonging to a cluster. The technique is iterative and all the neighbor points are iterated until no new neighbors are found. These set of points are then declared as a cluster. The process is repeated for all the points and designated as either a noise or a cluster point. In case two clusters are close (i.e, the distance between two border points from two clusters is less than ϵ), the clusters are merged into one.

The single molecule parameters (such as, density of single molecules, the cluster spread/area, number of molecules per cluster etc.) are critical indicators of cell physiology , and have deep implications in cell and disease biology. In practice, single molecule protocols from gene expression to cell transfection and microscope preparation to single molecule data analysis are critical to carry out diverse studies in Biology and Medicine.

References

1. Tsien RY (1998) The green fluorescent protein. Annu Rev Biochem 67:509–544
2. Jinek M et al (2012) A Programmable dual-RNA-guided DNA endonuclease in adaptive bacterial immunity. Science 337:816–821
3. Wang JY, Doudna JA (2023) CRISPR technology: a decade of genome editing is only the beginning. Science 379:251
4. Aravinth S, Varghese JM, Pant N, Mondal PP (2025) Super-resolution optical microscopy reveals accumulation of photoactivable dengue protein (Dendra2-NS2B) in the endoplasmic reticulum. Sci Rep 15:10358
5. Varghese JM, Joshi P, Aravinth S, Mondal PP (2022) Single-molecule super-resolution microscopy reveals formation of NS2B3 protein clusters on mitochondrial network leading to its fragmentation during the onset of dengue (Denv-2) viral infection, BioRxiv, 2022. https://doi.org/10.1101/2022.12.18.520514

6. Addgene. https://www.addgene.org/
7. Addgene plasmid #115906. http://n2t.net/addgene:115906. RRID :Addgene 115906
8. addgene plasmid #11910. http://n2t.net/addgene:11910. RRID:Addgene11910
9. Addgene plasmid #57483. http://n2t.net/addgene:57483. RRID:Addgene57483
10. Gudheti MV et al (2013) Actin mediates the nanoscale membrane organization of the clustered membrane protein influenza hemagglutinin. Biophys J 104:2182–2192

Stimulated Emission Depletion Microscopy (STED) and Its Single Molecule Analogues

11

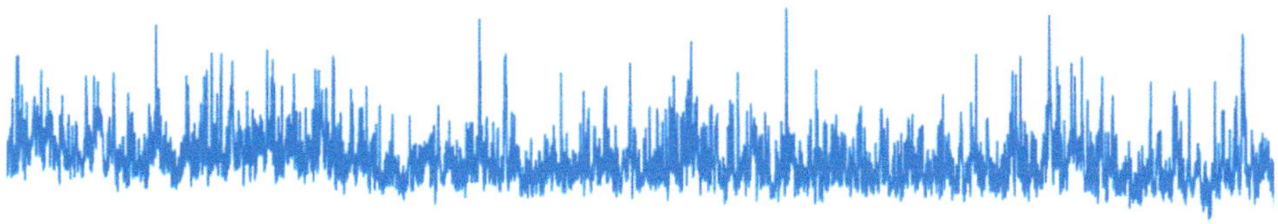

Stimulated emission depletion (STED) is a powerful super-resolution technique that allows visualization of sub-diffraction features and biological structures well beyond the diffraction limit ($\sim\lambda/2$) set by classical light. Traditional light microscopes are limited by the diffraction of light, which prevents them from resolving features smaller than this limit. STED microscopy overcomes this by using a dual-beam illumination, with a regular Gaussian beam to excite ("switch-on") molecules emitting fluorescence in a diffraction-limited Gaussian spot and a doughnut beam to "switch-off" or quench molecules at the periphery of the spot. This allows controlled switching OFF of molecules thereby shrinking the system PSF to as small as possible (theoretically), thereby improving the resolution by many-fold. Over the years, the technique has advanced beyond traditional STED, and many variants (two-photon STED, time-gated STED, MINSTED, MINFLUX) were invented that improved resolution and provided additional features (3D imaging, depth imaging, and multi-organelle imaging). This has pushed PArticle Resolution shift (PAR shift) towards the actual size of a single molecule. PAR-shift achieved by STED-based microscopy techniques has enabled the study of protein-protein interaction, visualization of biological structures, and nanoscale processes in living cells in real-time.

This chapter discusses a distinct concept for realizing super-resolution microscopy. The technique called STED came to focus in the early 90s and has undergone several iterations and developments over the years. We will begin with the basic idea of STED microscopy and build upon its advanced variants. These variants have realized the true potential of

P. P. Mondal, S. Hess, *Single-Molecule Super-Resolution Optical Microscopy*,
https://doi.org/10.1007/978-3-032-13442-4_11

STED microscopy and have demonstrated a PAR-shift approaching the actual size of a single molecule.

The root of STED concept goes back to 1990, when it was shown for the first time that two states (ON/OFF) of a molecule may be created adjacent to each other, even at a distance much closer than the diffraction limit ($\sim\lambda/2$) [1–3]. The following years have seen STED improving steadily and gaining popularity. A big improvement came with the development of three dimensional STED or $4pi - STED$, making it a complete imaging system [4]. The technique combines two independent concepts, with STED for stimulated emission depletion in the lateral plane and 4Pi for high-resolution axial scanning. Here, 4pi improves the axial resolution, whereas STED improves lateral resolution. Subsequent improvements led to two-photon STED. This combines the strength of two methods, two-photon excitation for better depth penetration and STED for high spatial resolution in a scattering sample [5]. Another advance is time-gated STED, that uses time-gated detection to further extend the resolution of STED microscopy [6]. The technique works by selectively detecting fluorescence after a delay following excitation, resulting in a strong reduction of the background and improvement of the signal-to-noise ratio.

Recently, MINSTED and MINFLUX techniques have been introduced that use single molecule switching along with the STED concept to entail PAR-shift close to sub-nanometer resolution [7, 8]. MINFLUX and MINSTED essentially combine the strengths of both STED and SMLM (fPALM/PALM/STORM). The technique uses the STED concept to search for the point of minimal STED and uses the STED doughnut as a movable reference coordinate for near exact fluorophore localization.

Although STED is impressive, it requires fluorophores that are capable of switching ON/OFF rapidly and are compatible with biological specimens. A list of these fluorophores is tabulated at the end of this chapter. We will begin the chapter with a basic understanding of the STED concept before pounding on practical aspects and key variants.

11.1 The STED Concept

The basic concept behind STED microscopy is stimulated emission, where the interaction of an excited state fluorophore with an incoming incident photon of appropriate energy causes its relaxation to the ground state. Notably, this is a key relaxation process in modern-day LASERs that involves light amplification by stimulated emission. In STED microscopy, this translates to the forced silencing of selective fluorescent single molecules by stimulated emission.

The STED microscopy involves two major states : *State B* - This is the excited state of fluorescent molecule that can be accessed by exposing it to an external excitation light (of specific wavelength). This state is also called the ON or fluorescence state *State A* - This is the lower energy level state to which the molecules relaxes giving out a light of higher wavelength or lower energy. This state is also called OFF state or ground state. The process begins with all the molecules in excited state which can be induced by an external

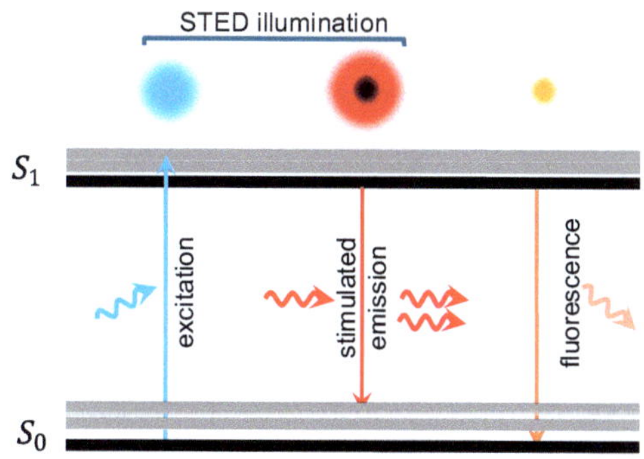

Fig. 11.1 The schematic diagram stating the principle behind STED microscopy. Alongside the regulator Gaussian excitation beam and STED beams are also displayed

radiation source of appropriate wavelength. This is as shown in the Jablonski energy diagram Fig. 11.1. Upon illumination by a light of appropriate energy, the fluorophore absorbs a photon and makes a transition to the excited state (S_1). This is practically achieved by a Gaussian beam that excites all the molecules falling in the Gaussian profile. The fluorophores returns to the ground state by the process of spontaneous emission (natural decay) which typically takes few nanoseconds. However, this can be avoided by exposing the molecule to a second light source of lower energy to cause de-excitation by the process of stimulated emission thereby emitting a stimulated photon. (see, Fig. 11.1). The stimulated photon has the same characteristics (wavelength) of that of light causing de-excitation, thereby distinguishing it from spontaneous fluorescence emission. The probability of this process is called *depletion*.

The mechanism behind the excitation and depletion of a fluorophore can be understood based on a two level system. The molecule is assumed to be in two distinct states: *state A* and, *state B* with distinct energy levels. The transition from one to the other state can be driven by light at a rate given by,

$$k_{A\to B} = \sigma_{A\to B}\, I(x) \tag{11.1}$$

where $I(x)$ is the intensity at a given location x within Gaussian beam profile, and $\sigma_{A\to B}$ is the molecular cross-section for $A \to B$ transition.

Assuming that the state B is long lived and the transition $A \to B$ commences at $t = 0$, then the probability of finding the molecule in the state A after $t > 0$ is,

$$P_A(x) = e^{-k_{A\to B}\, t} = e^{-\sigma\, I(x)\, t} \tag{11.2}$$

Now, consider that the transition is caused by a standing wave i.e, $I(x) = I_{max}\, cos^2(kx) - I_{max}\, cos^2(2\pi x/\lambda)$. So, the probability for transition becomes,

$$P_A(x) = e^{-\sigma I_{max}\, cos^2(2\pi x/\lambda)\, t} \tag{11.3}$$

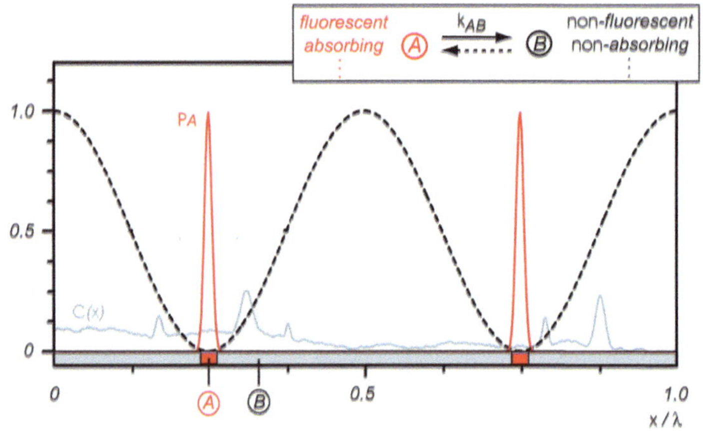

Fig. 11.2 A standing wave (of wavelength λ) for photoswitching molecules from one to another state. The corresponding probability $P(x)$ of finding the molecule in A is confined to nodal points. Image is taken from, Appl. Phys. A 77, 859–860 (2003)

The above equation (Eq. 11.3) states that, $P_A = 1$ at the intensity nodes i.e, $x_i = n\lambda/4$, where n is odd. This says that the molecule is in state A. On the other hand, $P_A \rightarrow 0$ for $x \neq x_i$ (n even). This means that *state B* can be established everywhere except for the very point with zero intensity i.e, x_i (see, Fig. 11.2). So, two distinct states can be created in close proximity i.e, with molecular proximity [3].

11.2 The Basic STED Microscopy

STED microscopy uses the principle of emission and forced depletion of molecules to achieve high resolution. Here we will provide a simplistic theory of STED in a very idealistic situation [9].

11.2.1 Theory behind STED Microscopy

Consider a two-level molecular system, with two energy states A and B (say, ground and excited state or conformational and isomeric states) between which the molecule can be transferred. Light of appropriate wavelength is used for the transition $A \rightarrow B$, however the transition $B \rightarrow A$ may be spontaneous or stimulated by light or any other means.

Let the transition rate be denoted by k_{AB} and k_{BA} for the transitions $A \rightarrow B$ and $B \rightarrow A$, respectively. Accordingly, the changes in the normalized population can be expressed as,

$$\frac{\partial}{\partial t}N_A = -\frac{\partial}{\partial t}N_B = -k_{AB}N_A + k_{BA}N_B \tag{11.4}$$

Assuming normalized population, $N_A + N_B = 1$, and under steady state condition i.e, $\frac{\partial}{\partial t}N_A = -\frac{\partial}{\partial t}N_B = 0$ gives,

$$\begin{cases} N_A = \frac{k_{BA}}{k_{AB}+k_{BA}} \\ \\ N_B = 1 - N_A \end{cases} \tag{11.5}$$

So, the above set of equations denote the equilibrium population of states A and B, respectively.

Now, we are interested in depleting state A which is achieved by light via the transition $A \rightarrow B$, for which the rate is directly proportional to the incident photon flux (I) i.e,

$$k_{AB} = \sigma \ I \tag{11.6}$$

where, σ is the molecular cross-section.

Incorporation of Eq. 11.6 produces the following expression for the equilibrium population of state A,

$$N_A = \frac{k_{BA}}{(\sigma \ I + k_{BA})} = \frac{k_{BA}/\sigma}{(I + k_{BA}/\sigma)} \tag{11.7}$$

Note that if $I >> I_{sat} = k_{BA}/\sigma$, then $N_A \rightarrow 0$ i.e, all molecules end up in B state. Here, I_{sat} is referred to as the saturation intensity.

Assuming a spatial intensity distribution $I(\vec{r})$ which is much greater than I_{sat} with zero at the center ($\vec{r}_0$) ensures that all molecules end up in the state B except those at $\vec{r}_0$. Expanding the intensity function we get, $I(\vec{r}) = I_{max} \ f(\vec{r})$, where $f(\vec{r})$ is a diffraction-limited spatial function with $f(\vec{r}_0) = 0$ at the center. Note that, the molecules found in state A squeezes to a point. This create sharp regions (at the center) of molecules in state A.

For a standing wave, $f(x) = sin^2(2\pi x/\lambda)$, the central region with molecules in state A has a FWHM of,

$$\Delta x = \frac{\lambda}{\pi} \ sin^{-1}\left\{\sqrt{\frac{k_{BA}}{\sigma \ I_{max}}}\right\} \tag{11.8}$$

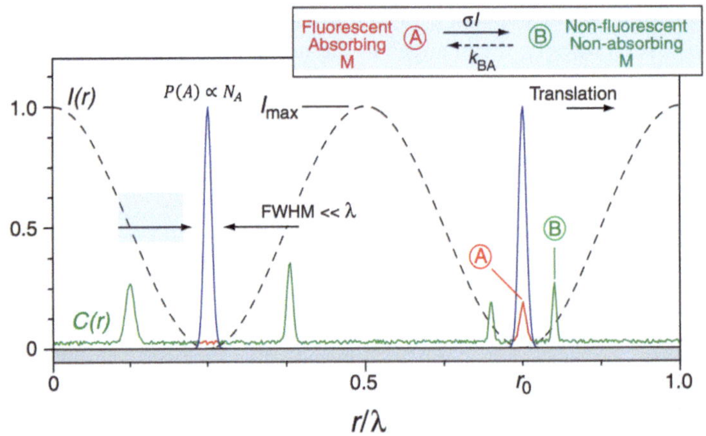

Fig. 11.3 The photoswitching of molecules is caused by a standing wave of intensity $I(r)$ and wavelength λ. The probability of finding molecules in state A is, $P(A) \propto N_A$, and are confined to the nodal points (blue). Here, $C(r)$ represents the sample. Image is taken from, Nature Biotechnology 21, 1347–1355 (2003)

Noting that, $k_{BA}/\sigma = I_{sat}$, and using Taylor series expansion for the sin^{-1}-term (neglecting higher orders), we get,

$$\Delta x \approx \frac{\lambda}{\pi} \sqrt{I_{sat}/I_{max}} = \frac{\lambda}{\pi \sqrt{\zeta}} \tag{11.9}$$

where, $\zeta = I_{max}/I_{sat}$ is the saturation factor. As an example, for $\zeta = 10{,}000$, gives $\Delta x \approx \lambda/100$. Thus by increasing the intensity I_{max}, one can squeeze the spot with A molecules (see, Fig. 11.3).

In Fig. 11.3, if state A (not B) is involved in fluorescence, the signal emerges from the narrow region defined by N_A (see, Fig. 11.3). It is the center spot where the probability of finding molecules in state A is maximum i.e, $P_A(r_0) = 1$. This simple observation allows fluorescence imaging with resolution far below the diffraction limit. In practice, several nodes are scanned across the sample $C(r)$ for super-resolving the specimen.

11.2.2 Optical Setup of a Standard STED System

STED may be thought of as a technique that belongs to PSF engineering technique, but unlike other techniques this is achieved by manipulating the ON/OFF state of a molecule. The technique requires two light sources: first, the excitation beam to excite the target molecules, second the STED beam to deplete molecules in regions where fluorescence is not desired. Both the beams are super-imposed to reduce the effectively PSF much below the diffraction limit ($\sim\lambda/2$) thereby super-resolving features in the specimen.

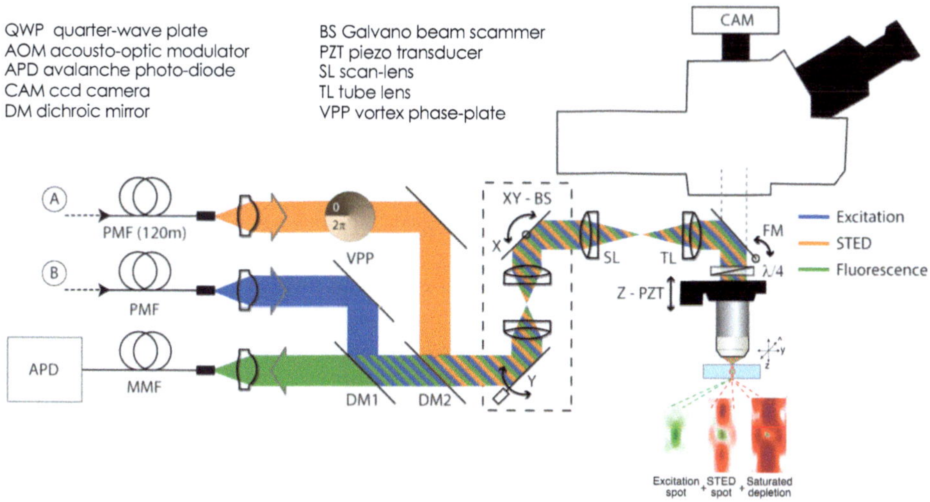

Fig. 11.4 Optical setup of a standard STED microscopy system. The image is adapted from, S. Berning et al., Nanoscopy in a Living Mouse Brain, Science 335, 551 (2012); S. W. Hell, Toward fluorescence nanoscopy, Nature Biotech. 21, 1347–1355 (2012)

A standard point-scanning STED microscopy system is shown in Fig. 11.4. Both excitation and STED are accomplished with synchronized laser pulses. Polarization maintaining single-mode optical fiber (PMF) are used to stretch the beam pulses to hundreds of picosecond. A lens is used to collimate the beams. While excitation beam (blue) is a standard Gaussian beam, the STED beam (A) is a donut-shaped beam. The donut shape is realized using a vortex phase-plate/phase mask that shifts the phase of the cross-section of the beam from $0 - 2\pi$ in a spatially dependent manner as shown in Fig. 11.4 (orange beam). Alternately, spatial light modulator can be used to achieve donut intensity distribution as well. It is important to have correct polarization to ensure matching the circular polarization and chirality of the vortex phase mask. This generates the desired intensity distribution i.e, central intensity zero by cancelling intensity contributions at the beam centre. Subsequent to the creation of donut cross-section beam shape, it is combined with the Gaussian beam using a dichroic mirror (DM2). The combined beam is then focussed to a pair of relay lenses to the XY-scanning mirror (XY-BS and FM). A pair of lenses (SL and TL) are also used along with the scanning mirrors. The scanning mirror enables scanning of the entire specimen in a point-by-point manner just like confocal system. Finally, the beams are focussed by a high NA objective lens (placed on a Z-piezo-stage) on to the sample. The regular excitation spot (Gaussian cross-section along XY) shown by Green is overlapped with the STED donut-shapped spot (red). The saturation depletion caused by the STED beam reduces the fluorescence at the periphery, except at the zero point leaving a much smaller (sub-diffraction size) spot. This is way smaller than the diffraction-limited confocal spot. The emitted fluorescence (represented by green color) is then collected by the objective lens and follows the path back until it reaches

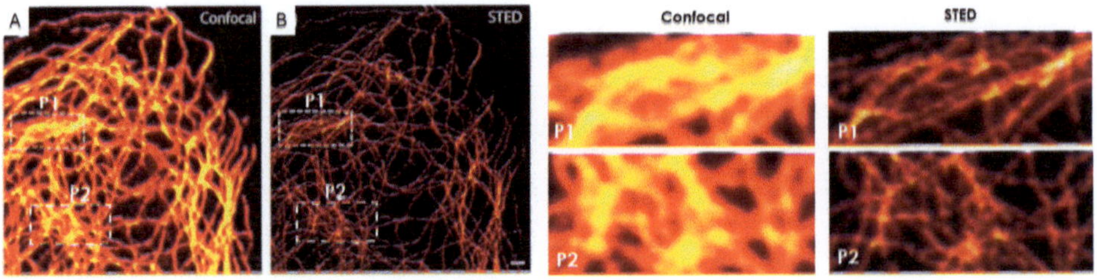

Fig. 11.5 Single color STED and confocal images of mitotubular network of Ptk2 cells. The images are taken from, D. Wildanger et al., A STED microscope aligned by design, Optics Exp. 17, 16100–16110 (2009)

the dichroic mirrors (DM1 and DM2) that allow it to pass through to the avalanche photo diode (APD) via a multimode fiber. To reconstruct the entire XY plane of the specimen, the scanning mirrors are used that sample the specimen with a distance half of Rayleigh criterion i.e, $\sim\Delta_{psf}^{sted}/2$ (ideally). Since, $\Delta_{psf}^{sted} << \Delta_{psf}^{conf}$ this allows finer scanning of the entire specimen as compared to confocal microscopy. In principle, the spot size is not limited but by the quality of the naught at the centre and the saturation factor of depletion. Finally, the piezo (Z-PZT) can be moved in small steps (greater than PSF along Z) and all the z-planes can be scanned to obtain the entire volume (provided the sample is transparent enough).

Figure 11.5 shows a single-beam STED super-resolved image of a microtubular network of PtK2 cells. Alongside confocal images are also shown for comparison. It is evident that tangled network structures are better resolved as compared to diffraction-limited confocal microscopy. This is better evident in the enlarged sections (P1 and P2) of the image.

Just like multicolor confocal, STED microscopy can also be operated in multicolor mode. This require more lasers for excitation of STED dyes and more detection channels with additional optics. Figure 11.6 shows two color and four color super-resolved STED images of biological specimens. The two-color images shown here are immunolabeled fixed cells and immunolabeled subunits in amphibian NPC. The four color images are also that of immunolabeled fixed cells [10].

11.3 Variants of STED Microscopy

STED microscopy has evolved over the last 3 decades since its inception in 1990s. Since then, a number of important variants have emerged that have found applications in a large spectrum of studies and bettered our understanding of sub-cellular organelles and functions therein.

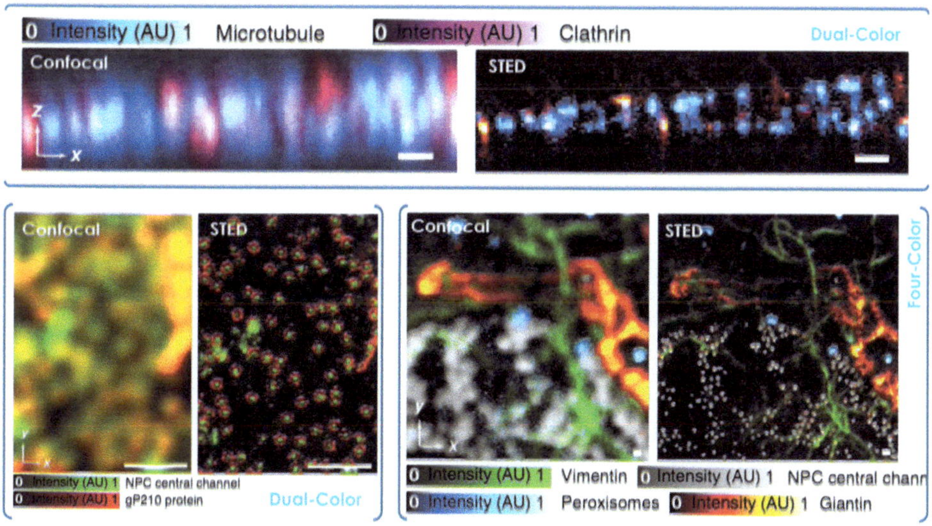

Fig. 11.6 Multi-color STED and confocal images of subunits in amphibian NPC. The images are taken from, G. Vicidomini et al., sted super-resolved microscopy, Nature Meth. 15, 173 (2018)

11.3.1 Three Dimensional STED

3D imaging is at the core of any imaging system that makes it complete and better realize its full potential. We noted that general STED microscopy provides enhanced resolution in the lateral XY plane. So, to gain in z-resolution the PSF needs to be engineered such that zero intensity appears in all directions with high intensity surrounding it. This can be achieved by super-imposing two incoherent beams, one producing the regular donut-shapped beam (along XY) and the other producing bottle-shapped beam (along Z). Both the beam types along with a regular Gaussian beam is shown in Fig. 11.7. Technically, SLM is used to combine these two depletion patterns. This has the added advantage of aberration correction and beam alignment.

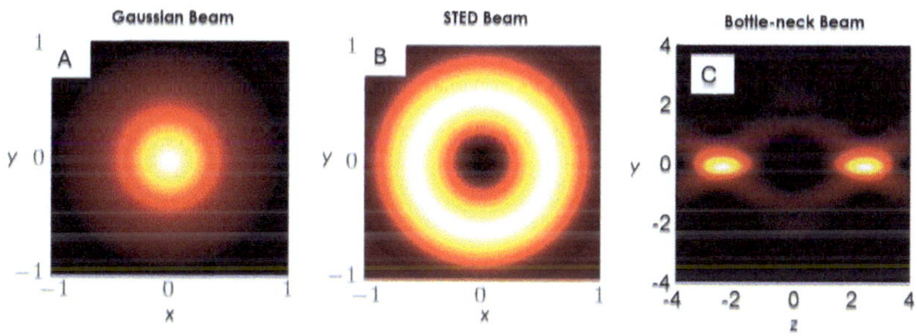

Fig. 11.7 Regular Gaussian, donut and bottle-neck beams

Another popular technique for realizing 3D STED imaging is called *4Pi-isoSTED* [9]. This microscopy technique uses $4Pi$ optical configuration where two opposing coherent beams (derived from the same source) are superimposed. This results in interference of both the beams, providing isotropic 3D PSF. The downside is however the fact that changes in the refractive index of the specimen affect the interference and effective PSF, thereby strongly limiting its performance to transparent thin samples only, such as mouse/human cell.

11.3.2 Two-photon Excitation STED

A natural extension of STED microscopy is two-photon STED. This integrates the benefits of two-photon excitation necessary for depth-penetration and STED for lateral super-resolution. The technique has a strong presence in neuroimaging where scattering is dominant [11]. This has the advantage of an in-depth imaging of neuronal structures and function within neural tissue as compared to conventional imaging tool. Two-photon STED is a perfect imaging tool for probing brain function with a resolution much beyond the diffraction limit. Unlike single photon excitation, two photon excitation (2PE) is a rare event. The excitation requires absorption of two low energy photons almost instantaneously (with a time-interval of few femto-seconds) (see, Fig. 11.8b). So, 2PE requires ultra-high laser peak power ($\sim$few GW/cm^2) that is normally delivered by femtosecond laser. To realize 2PE in a STED configuration, pulsed STED is preferred rather than continuous STED. In a pulsed STED system, two synchronized laser pulses

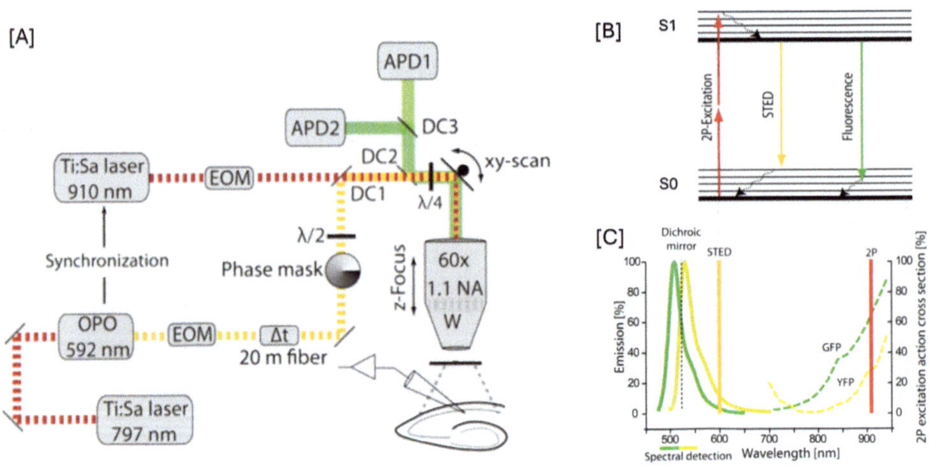

Fig. 11.8 (**a**) Schematic diagram of a two-color two-photon STED microscope (**b**) Jablonski diagram of 2PE-STED. (**c**) Spectral distribution along with other details. Images taken from, P. Bethge et al., Two-Photon Excitation STED Microscopy in Two Colors in Acute Brain Slices, Biophys. Jl. 104, 778–785 (2013)

are used, the first pulse excites the fluorophore followed by a second red-shifted pulse that de-excites fluorophores in the peripheral region.

Figure 11.8 shows the schematic diagram of a 2PE STED system. Note that an additional APD is used for multicolor imaging. Two pulsed lasers are used: $Ti : Sa$ 797 nm and $Ti : Sa$ 910 nm. Here, $Ti : Sa$ 797 nm laser is used in conjugation with an OPO to convert the wavelength to a 592 nm light for depletion, whereas $Ti : Sa$ 910 nm is used for two-photon excitation. Note that, the use of two pulsed lasers allow easy synchronization of pulses. Subsequently electro-optical modulator (EOM) is used in both the optical arms for controlling the laser intensities of two-photon and STED laser beams. In addition, this allows faster modulation of beams to avoid unwanted fluorescence excitation or photobleaching. The pulses from the $Ti : sa/OPO$ 592 nm are broadened by a 20 m long polarization-preserving single-mode fiber. Subsequently, a helical phase-mask is used to form the doughnut. Phase plates ($\lambda/2$ and $\lambda/4$) are used to optimize the doughnut quality (shape and null). Subsequently, the beams pass through a XY scan mirror and projected at the back focal plane of a high NA objective lens which focuses the light on to the specimen.

The emitted fluorescence was detected in the descanned mode and separated from the excitation and STED light by a long-pass dichroic mirror (DC2). Note that, a dedicated band-pass filter (593 ± 40 nm) was used to spectrally clean up the STED laser beam. Following the spectral clean-up, the fluorescence signal (spectral window, 480 – 550 nm) was spectrally divided into two channels by the dichroic mirror (DC3). Finally, the signal/light is focused onto multimode fiber which terminates on the avalanche photodiodes (APD1 and APD2). The spectral details of the fluorophores along with the 2PE and STED light is shown in Fig. 11.8c.

The excitation and STED PSF used for synchronized scanning of the specimen is shown in Fig. 11.8a. The fluorophores YFP and GFP are expressed in neurons and microglia, respectively (see, Fig. 11.9). The images shown in (B1–B2) are a merges of both the color channels, whereas B3 is linearly unmixed showing two separate channels (GFP (green) and YFP (red)). Here B1–B2 shows superresolved image of microglial process contacting dendritic spines, where dotted lines indicate spine neck widths. So, it is evident that, 2PE-STED super-resolution microscopy is a potential tool for neurobiology.

11.3.3 Time-Gated STED Microscopy

Another important variant of STED microscopy is when the microscope is operated in time-gated mode. In this mode, the fluorescence post STED beam is collected in a time-gated manner to avoid residual fluorescence to reach the detector. Such a technique helps in improving the signal-to-noise ratio for time-gate less than average lifetime and the overall resolution [12]. Specifically, in continuous mode, gating strongly suppresses low spatial frequencies and reduces the influence of local variations in fluorescence. Moreover, this takes care of uncorrelated background (ambient light, undue fluorescence by STED laser and scattering) in the detection gate.

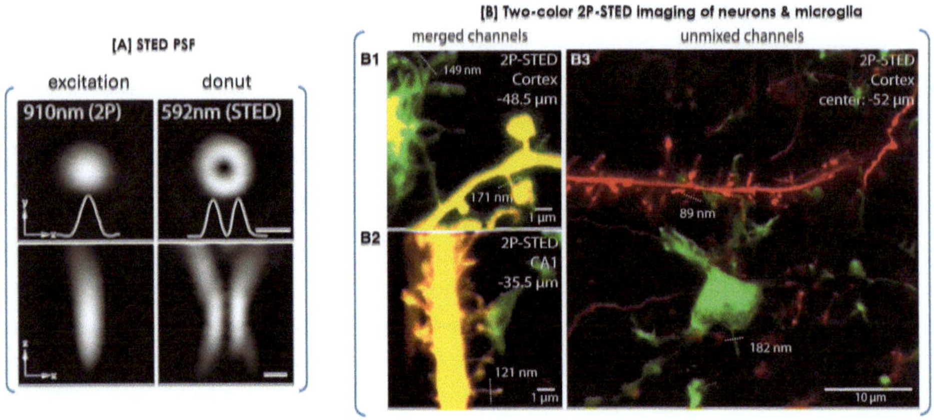

Fig. 11.9 Two-color 2PSTED imaging of labeled neurons and microglia. The images are taken from, P. Bethge et al., Two-Photon Excitation STED Microscopy in Two Colors in Acute Brain Slices, Biophys. Jl. 104, 778–785 (2013)

In a pulsed STED scheme, the fluorescence photons should be detected just after the STED pulse has left [13]. Note that the obtainable resolution scales with the duration of depletion beam. So by applying pulsed excitation coupled with time-gated detection helps in reducing the residual fluorescence from the slopes of STED beam by ensuring fluorescence from those molecules that have been exposed to the beam for a duration $> T_g$ (detection delay) after excitation [14, 15].

Consider a two-level molecule (with S_0 and S_1 as the ground and excited singlet states) undergoing excitation depletion process. At time $t = 0$, the molecules are assumed to be in S_1 state after a brief excitation pulse. In addition, we assume that the pulse-repetition time ($T = 1/f$) i.e, the time between two pulses is longer than the excited-state lifetime τ of the molecule. In this scenario, the rate equations for the population (P_{S1}) of first excited state is given by [12],

$$\frac{\partial P_{S1}}{\partial t} = -k_{S1} P_{S1} - k_{STED} P_{S1} \tag{11.10}$$

where, $k_{S1} = 1/\tau$ and $k_{STED} = \sigma_{STED}\ I_{STED}$ are the rates for spontaneous and stimulated emission. Here, τ is the excited state lifetime, σ_{STED} is the stimulated state cross-section, and I_{STED} is the transient STED intensity.

Noting that $P_{S1}(0) = 1$, the fluorescence emission rate at a time t post excitation is,

$$F(t) \propto P_{S1} = q_{fl}\ k_{S1}\ e^{-k_{S1}t}\ e^{-k_{S1}\zeta^* t} \tag{11.11}$$

where, q_{fl} is the fluoresence quantum yield, and $\zeta^* = I_{STED}/I_S = k_{STED}/k_{S1}$ as the saturation factor. I_{STED} and $I_S = k_{S1}/\sigma_{STED}$ are the STED and saturation intensity, respectively. Here, the first and the second term represents the spontaneous decay and stimulated decay (due to STED beam), respectively.

Close to the zero-intensity point ($r = 0$), the STED intensity profile (doughnut) can be approximated by a parabola i.e, $I_{STED}(r) \approx 4\ I_{STED}\ a^2r^2$, where a is a constant. Using this relation, the PSF can be expressed as [12],

$$h(t,r) \approx \begin{cases} h_c(r)\ e^{-k_{S1}t}\ e^{-4a^2r^2\zeta^* k_{S1}t}, & for\ 0 \le t \le T_{STED} \\ h(T_{STED,r})\ e^{-k_{S1}t}\ for\ T_{STED} \le t \le T \end{cases} \tag{11.12}$$

Noting that the fluorescence is detected for $t > T_g$ after the excitation pulse, the PSF with time-gated detection (for $T_g \ge T_{STED}$) is given by [12],

$$h_P(r) = \int_{T_g}^{T} h(t,r)\,dt \cong h_c(r)\ e^{-T_g/\tau}\ e^{-4a^2r^2\zeta^* T_{STED}/\tau} \tag{11.13}$$

We have assumed that $T_g >> \tau$ making the term $e^{-T_g/\tau}$ negligible, producing,

$$h_P(r) = \int_{T_g}^{T} h(t,r)\,dt \cong h_c(r)\ e^{-4a^2r^2\zeta^* T_{STED}/\tau} \tag{11.14}$$

This resembles a Gaussian distribution for which the full-width-at-half-maximum (FWHM) is given by,

$$FWHM = d_c/\sqrt{1 + d_c^2a^2\zeta} \tag{11.15}$$

where, d_c is the FWHM of a Gaussian distribution, $\zeta = \zeta^*\ T_{STED}/(\tau\ ln2) = T_{STED}\ I_{STED}\sigma_{STED}/ln2$ is the saturation factor. It is evident from the above expression that the FWHM depends on the pulse energy ($T_{STED}I^*_{STED}$) and cross-section, and not on the fluorescence lifetime. Thus it is quite apparent that, in the pulsed STED technique, the collection of photons immediately after the STED pulse ($T_g = T_{STED}$) produces the sharpest PSF. However, the impact of time-gating is negligible if the pulse-width of the STED beam T_{STED} is shorter than excited state lifetime τ.

Figure 11.10 shows the PSF and images obtained using gated pulsed STED (gps-STED/ g-STED). Both theoretical PSF (using Eq. 11.14–11.15) and experimental PSFs are shown. The experimental PSF measured with fluorescent nitrogen vacancy (NV) color centers in diamond (see, Fig. 11.10a). Alongside confocal, continous wave STED (CW-STED) and non-gated pulsed PSFs (ps-STED) are also shown. In addition, gated pulsed STED (gP-STED) images of microtubule of fixed PtK2 cells (immunolabellled with dyes, KK114 and ATTO647N) are also shown. The improvement over confocal and between gated and non-gated STED is quite apparent. The images are adapted from Refs. [12] and [15].

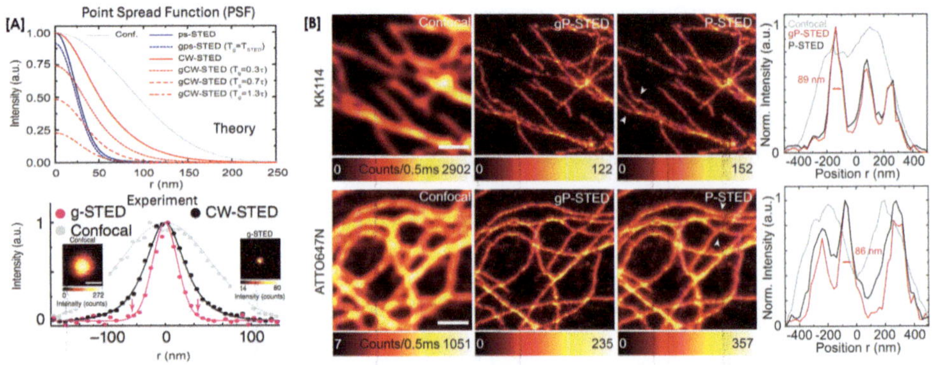

Fig. 11.10 System PSF and time-gated STED reconstructed images of microtubule of PtK2 cells. Images taken from, G. Vicidomini et al., STED Nanoscopy with Time-Gated Detection: Theoretical and Experimental Aspects, PLoS One 8, e54421 (2013)

11.3.4 MINSTED and MINFLUX

STED microscopy works by confining the fluorescence to a sub-diffraction sized region around its central minimum. The fluorescence ability and the location of STED doughnut also describes the effective point spread function (PSF) of the STED microscope which is a Gaussian with a FWHM of,

$$FWHM = \frac{\lambda}{2NA\sqrt{1 + I/I_s}} \tag{11.16}$$

where, I and I_s are the focal peak intensity crest and the intensity at which fluorescence is half, respectively. At a conceptual level, detecting a single signal photon per fluorophore may render the location of single molecule which is perfectly defined by the doughnut.

The central theme of MINSTED is the ability to better the location of single molecule by minimal STED (MINSTED) effect. In MINSTED, the doughnut position is made identical to the location of the fluorophore coordinate so as to ensure minimal STED. In contrast, another technique called MINFLUX operates by searching for the position with minimal fluorescence (MINFLUX) using a doughnut excitation (laser beam with zero central intensity). Specifically for MINSTED, since the STED beam can be steered at sub-nanometer precision (using beam deflectors), the doughnut position can be used for finding the exact position of the single molecule in a sample.

MINSTED works in an iterative way to determine the actual position of the single molecule/fluorophore. First, the position of a single molecule is roughly identified by a continous-wave laser. Next the co-aligned excitation and STED beams were circled around the position estimate C_i with a radius $R_i = d_i/2$. Here R_i represents the size of effective PSF generated by co-aligned beams. Both C_i and R_i were updated after each photon detection. Note that, $i = 0$ corresponds to the diffraction-limited E-PSF diameter d_0 for

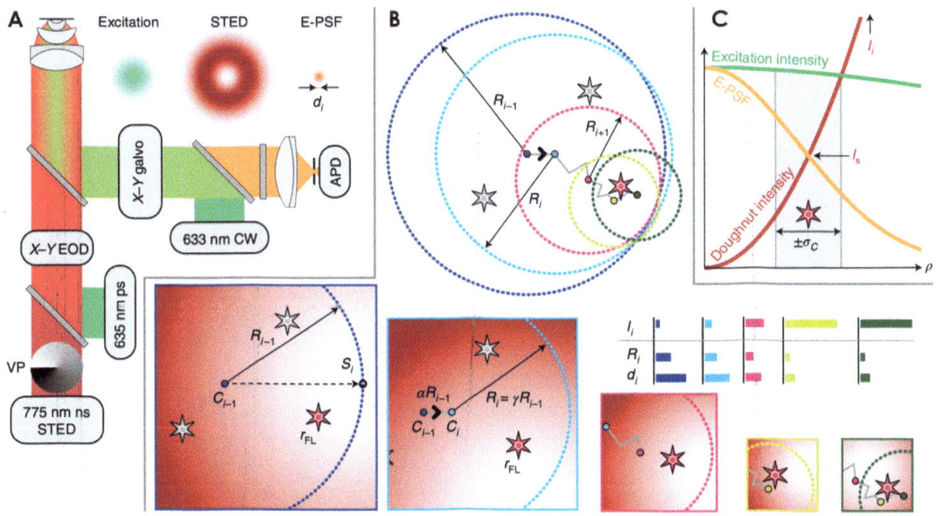

Fig. 11.11 MINFLUX, (**a**) The optical configuration of MINSTED. (**b**) The iterative process of finding the centroid and radius. (**c**) The effective PSF (E-PSF) and its variation with doughnut intensity I_{STED}. The figure is adapted from, Nature Photonics 15, 361–366 (2021)

STED intensity $I_{STED} = 0$ of co-aligned beams. Starting with $i = 0$, the scan centre C_i is shifted by a fraction (say, αR_i) towards the doughnut position (E-PSF maximum) when detecting the next photon (say, $i + 1$). In the next step, the doughnut intensity (I_s) is increased that reduces the radius R_i, thereby changing the centre and radius. Iteratively, the STED intensity is increased with each step in the direction of each detected photon. The whole process is schematically shown in Fig. 11.11 along with the E-PSF and doughnut intensity (I_s).

The iterative nature of MINSTED has shown a substantial improvement in resolution over existing standard STED microscopy. Stefan Hell's group has demonstrated this by imaging mitochondrial protein Mic60 in a U2 OS cells [7]. Fig. 11.12a–c shows the distribution of Mic60 in the mitochondria. The MINSTED image (Fig. 11.12c) shows ring like structures which are not apparent in standard STED images (Fig. 11.12c). For comparison confocal images are also shown (see, Fig. 11.12a). This is possible due to the sub-10nm resolution provided by MINSTED. This is in excellent agreement with the current understanding of Mic60 that are known to form small ring-like assemblies at the cristae junctions.

11.4 Fluorophores for STED

For realizing quality STED imaging, the single molecules/fluorophores must have specific photophysical properties. First, ideal separation of stimulated and regular fluorescence.

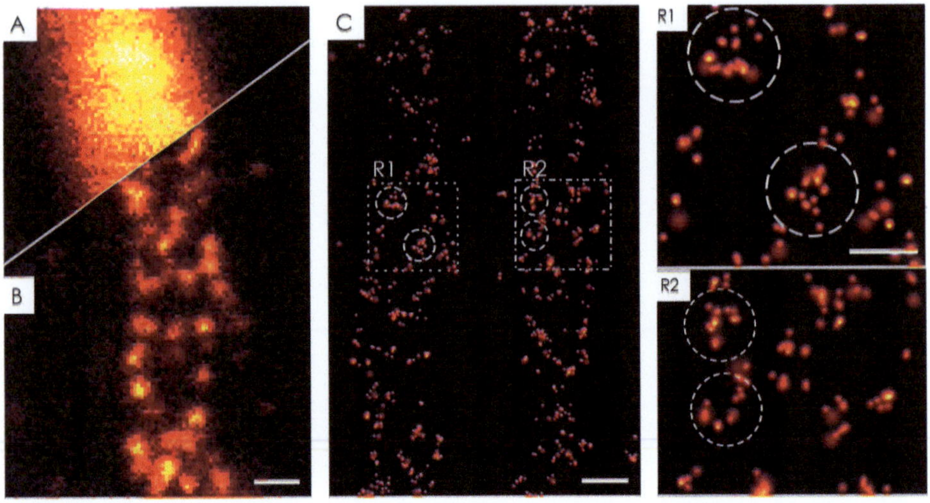

Fig. 11.12 Imaging Mic60 molecules in mitochondria, (**a**) Confocal image, (**b**) Standard STED image, and (**c**) MINSTED image. Alongside enlarged sections are also shown. Image adapted from, Nature Photonics 15, 361–366 (2021)

This means that the STED laser wavelength should be red-shifted, in specific the STED wavelength must not overlap with the absorption spectrum of the fluorophore. This avoids re-excitation, thereby reduces background and effective image resolution. Second, the stimulated emission cross-section (which is responsible for effective depletion efficiency) is directly proportional to the quantum yield (QY) and extinction coefficient (E_c) and inversely to the fluorophore lifetime. Fluorophores with high QY and E_c and a lifetime of few nanoseconds are optimal for STED. In addition, high photostability is a much needed property for quality STED microscopy.

One of the first fluorescent protein that have impacted biological research is Green Fluorescent Protein (GFP) and its engineered color variants. The protein is extensively used to study diverse biological processes both in live cells and multicellular organisms. The advantage of fluorescent proteins is their genetically encoded chromophore, and the specificity with which they can be labeled to the protein-of-interest or specific organelle. However, their limitations are, weak fluorescence, poor photostability, low quantum yield, and multimerization. Since high resolution STED require high intensity causing significant photobleaching , mostly fluorescent dyes are preferred. In specific, organic dyes are most suitable for STED microscopy which is due to their high brightness, and photostability. Some of the tried and tested dyes that have found widespread use in STED are appended in the table shown in Fig. 11.13.

Fluorophore	λ_{ex} (nm) ($\varepsilon \times 10^3$ $M^{-1}cm^{-1}$)	λ_{em} (nm) (Φ_{fl})	Brightness	τ_{fl} (ns)	Cell permeable	Recommended STED laser wavelength (nm)	Refs.
Long Stokes shift channel							
ATTO 425	439 (45.0)	485 (0.90)	40.5	3.6	–	577	46
CX (abberior LIVE 460L)	459 (28.0)	599 (0.55)	15.4	3.2	+	775	40
Chromeo 494	494 (55.0)	628 (0.07)	3.9	–	–	750	225
Green channel							
Alexa Fluor 488	495 (73.0)	519 (0.92)	67.2	4.1	–	590	226
510R (abberior LIVE 510)	496 (74.0)	520 (0.90)	66.6	3.8	+	590	227
MaP510	502 (74.0)	526 (0.68)	50.3	–	+	590	60
Yellow channel							
JF549	549 (101.0)	571 (0.88)	88.9	3.8	+	660	228
TMR (abberior LIVE 550)	555 (81.3)	580 (0.39)	31.7	2.1	+	660, 775	94
MaP555 (SPY555)	553 (88.0)	580 (0.40)	35.2	–	+	660, 775	60
JF571	571 (101.0)	590 (0.78)	–	–	+	775	104
Orange channel							
580CP (abberior LIVE 590)	585 (102.9)	609 (0.58)	59.7	3.6	+	775	229
JF585	585 (156.0)	609 (0.78)	121.7	–	+	775	230
ATTO 590	593 (120.0)	622 (0.80)	96.0	3.7	+	775	2,231
ATTO 594	603 (120.0)	626 (0.85)	102.0	3.9	–	775	232
Alexa Fluor 594	590 (73.0)	618 (0.66)	46.7	3.9	–	775	200
Red channel							
KK114 (abberior STAR RED)	638 (120.0)	655 (0.55)	66.0	3.4	–	775	233
abberior STAR 635P	638 (120.0)	651 (0.9)	–	3.3	–	775	232
SiR	645 (101.7)	661 (0.41)	41.7	2.7	+	775	51
JF646	646 (152.0)	664 (0.54)	82.1	–	+	775	228
ATTO 647N	646 (150.0)	664 (0.65)	97.5	3.5	+	775	225,234
JFX650	650 (150.0)	667 (0.53)	79.5	–	+	775	235
Far-red channel							
SiR700	689 (100.0)	716 (0.13)	13.0	1.4	+	810	236
MaP700 (SPY700)	700 (52.0)	720 (0.24)	12.5	–	+	810	60

ε, extinction coefficient; λ_{em}, emission maximum; λ_{ex}, excitation maximum; τ_{fl}, fluorescence lifetime; Φ_{fl}, fluorescence quantum yield; STED, stimulated emission depletion.

Fig. 11.13 Commonly used depletion-compatible organic dye molecules used in STED microscopy. The table is taken from, G. Lukinavicius et al., Stimulated emission depletion microscopy, Nat. Rev. Meth. Prim. 4, 56 (2024)

References

1. Hell SW, Wichmann J (1994) Breaking the diffraction resolution limit by stimulated emission: stimulated-emission-depletion fluorescence microscopy. Opt Lett 19:780
2. Hell SW, Kroug M (1995) Ground-state-depletion fluorscence microscopy: a concept for breaking the diffraction resolution limit. Appl Phys B 60:495
3. Hell SW, Jacobs S, Kastrup L (2003) Imaging and writing at the nanoscale with focused visible light through saturable optical transitions. Appl Phys A 77:859–860
4. Dyba M, Keller J, Hell SW (2005) Phase filter enhanced STED-4Pi fluorescence microscopy: theory and experiment. New J Phys 7:134
5. Moneron G, Hell SW (2009) Two-photon excitation STED microscopy. Opt Express 17:14567–14573
6. Vicidomini G et al (2011) Sharper low-power STED nanoscopy by time gating. Nat Methods 8:571–573
7. Weber M, Leutenegger M, Stoldt S, Jakobs S, Mihaila TS, Butkevich AN, Hell SW (2021) MINSTED fluorescence localization and nanoscopy. Nat Photon 15(5):361–366
8. Weber M et al (2023) MINSTED nanoscopy enters the Angstrom localization range. Nat Biotech 41:569–576

9. Hell SW (2003) Toward fluorescence nanoscopy. Nat Biotechnol 21:1347–1355
10. Vicidomini G et al (2018) STED super-resolved microscopy. Nat. Methods 15:173
11. Ding JB, Takasaki KT, Sabatini BL (2009) Supraresolution imaging in brain slices using stimulated-emission depletion two-photon laser scanning microscopy. Neuron 63:429–437
12. Vicidomini G et al (2013) STED nanoscopy with time-gated detection: theoretical and experimental aspects. PLoS One 8:e54421
13. Schrader M et al (1995) Monitoring the excited state of a dye in a microscope by stimulated emission. Bioimaging 3:147–153
14. Moffitt JR et al (2011) Time-gating improves the spatial resolution of STED microscopy. Opt Exp 19:4242–4254
15. Vicidomini G, Moneron G et al (2011) Sharper low-power STED nanoscopy by time gating. Nat Methods 8:571–573

12 Selected Topics in Single Molecule Biophysics

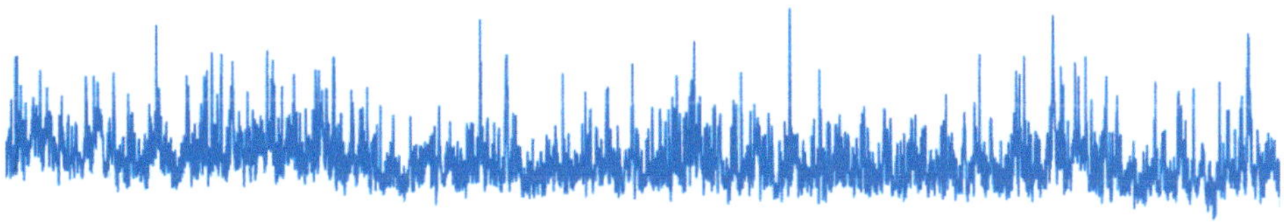

In this chapter, we explore key techniques that are intertwined with single-molecule imaging but may not be directly linked. Nonetheless, these techniques are key to understanding the behavior and characteristics of single molecules in living cells. Although visiting all methods is not possible and may not be in the scope of this book, we covered a few widely used ones. The techniques include single-molecule tracking and widely used single-molecule force spectroscopy techniques (Optical tweezers and Atomic force microscopy). In addition, we discussed a recent advance in SMLM, which is purely based on events, i.e., change in photon emission from single molecules. This has an advantage over traditional SMLM, specifically, with the event-based SMLM capable of high-density localization, large PAR-shift, high spatio-temporal resolution, and volume imaging. Finally, the chapter summarizes potential applications that were thought impossible a decade before, including protein folding and molecular kinetics in a live cell.

Single-molecule techniques are key to understanding life processes at the fundamental single-molecule level. They form the basis for addressing real-life issues, such as deciphering molecular mechanisms in a diseased cell to access its therapeutic potential, discerning disease progression, uncovering fundamental aspects of cell and protein functioning, drug discovery, and many more. The fact that these processes and related mechanisms occur at a single molecule level makes these techniques so special. In this chapter, special attention is given to the measurement of sub-nanometer length scales and forces that are fundamental to biological processes. Noting that all the processes involved in cellular processes (cell

P. P. Mondal, S. Hess, *Single-Molecule Super-Resolution Optical Microscopy*,
https://doi.org/10.1007/978-3-032-13442-4_12

mortality to replication) are driven by molecular forces, the study of these forces and the means to do so becomes essential to the understanding of fundamental biological processes. The last two decades have seen a revolution in these techniques that enable the measurement of sub-picoNewton forces and displacements of sub-nanometers generated by single biomolecules (protein complexes and DNA).

The force measurement techniques concern the measurement of lengths in the order of magnitude 10^{-10}–10^{-4} m and forces in the order of 10^{-14}–10^{-8} N. These length scales are associated with single DNA, proteins, nucleic acids, and their folding kinetics. For example, the measurement of RNA polymerase advancing a single base pair (0.34 nm) along DNA [1], and mechanical disruption of covalent bonds (few nanoNewtons) [2]. Moreover, the nucleic acid folding kinetics is found to be in the order of sub-picoNewton force [3]. All this is made possible by single molecule force spectroscopy techniques.

Another aspect of utmost importance is the dynamics of single molecules and the ability to track them as they interact with other molecules. Since single molecule tracking (SMT) requires a relatively longer time, the probes need to be bright, photostable, and easier in their genetic encoding process [4]. An ideal probe allows individual biomolecules or viral particles to be followed upto a timescale of minutes. Such a study reveals the distinct pathways these biomolecules take to complete their task in the living system. Modern SMT methods are capable of much more than tracking, i.e., observing multiple species with a temporal resolution down to a few tens of microseconds [5, 6]. Some of the pioneering applications are, myosin V walking on the actin filament, cargo transport mechanisms through nuclear pore complex, compartmentalization of highly dynamic plasma membranes, and validating the drug efficacy, among others [7–10]. These studies indicate the potential applications in fundamental biology and analytical chemistry, with the resolution down to a single molecule.

Finally, from super-resolution imaging viewpoint, a recent advancement has been the advent of event-based SMLM [11, 12]. The technique rely on differential detection of incoming signal photons and have the added advantage of reducing dynamic background noise. Compared to traditional SMLM, which suffers from limited temporal resolution and fixed frame rate of acquisition, event-based SMLM is fast and capable of imaging densely packed single molecules. Noting that the random Poissonian nature of photon arrival time and the consequent change in the event detector, this information can be used to generate multiple realizations of PSF ('$tPSFs$') to better estimate its position. In addition, the negligible read-time of the event detector substantially improves temporal resolution. Test on Dendra2-HA molecules in transfected NIH3T3 cells shows a position localization with a precision of ~6 nm, which is approximately 6.5-fold improvement over standard SMLM. This translates to a spatio-temporal PArticle Resolution (PAR) of $(2.3l_p \times 3\tau)$ of 42.35 par (where 1 par $= 10^{-11}$ meter.second) for event-based SMLM. However, brighter probes (such as Cy3) are capable of localization precision (l_p) of <5 nm at a collection time of $\tau = 3$ msec, giving a substantial improvement in spatio-temporal resolution (~12.70 par).

All these techniques form the basis of single-molecule imaging and aid single-molecule analysis. In this chapter, we explore techniques and methods that directly relate to single-molecule. In addition, we discuss a few potential applications of these techniques as well.

12.1 Single Molecule Tracking

As the name suggests, single particle tracking (SPT) is an imaging method to track the motion of a particle or single molecule in a biological system (such as a living cell). The technique is a unique tool to characterize the behavior of single protein molecules (interaction with other molecules, binding-unbinding behavior, and temporal dynamics) and observe their kinetics as they navigate through the cellular environment. The technique requires selecting a commercial microscope (widefield or scanning) and choosing a compatible fluorophore (bright and stable). In the simplest form of SPT, a widefield microscope with a collimated beam is used to illuminate the entire field of view. This is succeeded by following the particle throughout the observation time. Finally, the position of the molecule is obtained in time, and high-quality tracks are formed by linking emitter positions.

The SPT technology is rapidly changing and getting upgraded. Here, we choose to discuss a recent SPT technique that is simple and efficient [13]. The general outline of the system, along with the processes involved, is shown in Fig. 12.1a, b. The system consists of a 2D electro-optic deflector and a tunable acoustic gradient (TAG) lens. This combination is used in the illumination sub-system to rapidly scan (lateral XY-scan rate $\sim$50 kHz, and axial Z-scan of 70 kHz) using a focused laser spot (1 μm $\times$ 1 μm $\times$ 2 μm) in a predetermined 3D pattern. The photons emitted by the particle/single molecule are detected by an avalanche photodiode (APD) and converted to estimate the particle position. The processes typically involve four steps: (1) laser scanning in 3D, (2) illuminating and exciting the molecule, (3) real-time computation of the molecule's X, Y, Z position, and (4) finally, relocating the molecule. The final step is accomplished by using a PID feedback loop to drive the piezoelectric nanopositioner stage to re-center the molecule in the laser scan. The change in coordinates of the stage determines the new X, Y, Z position of the molecule. To carry the process in real-time (every 20 μs), a field-programmable gate array (FPGA) with assumed Gaussian density is employed. Fig. 12.1c shows the trajectory of a single Atto 647N molecule tracked for 160 seconds. Alongside, the intensity trace recorded during the trajectory of a single molecule (showing the typical signature of a single molecule) is also displayed (see Fig. 12.1d). Post localizing the emitters, tracks are reconstructed from these positions (see Fig. 12.1c). The technique works well at low emitter densities and is implemented using the nearest neighbor approach [14]. However, SPT becomes challenging at high emitter densities.

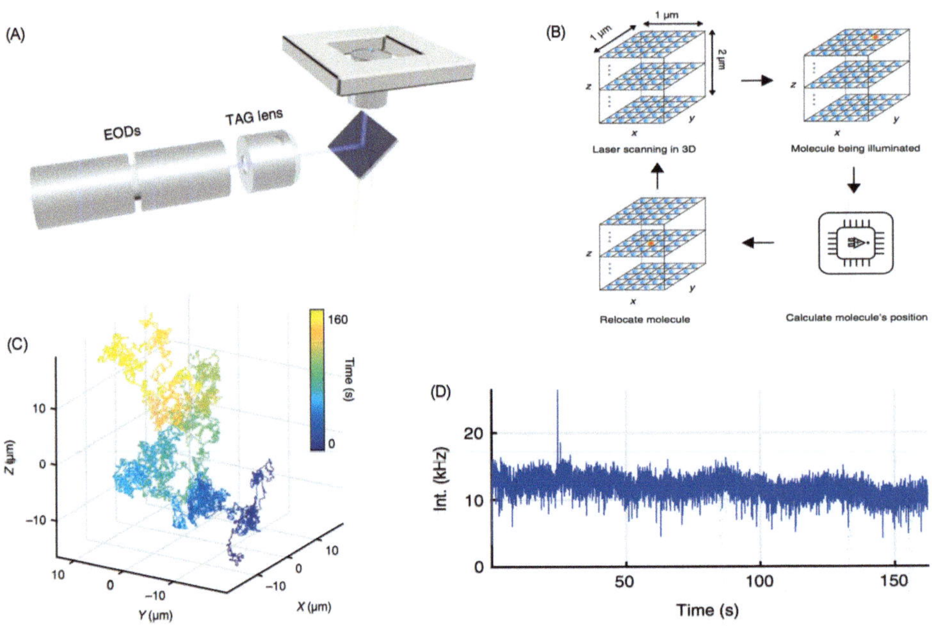

Fig. 12.1 (**a**) Schematic of a typical single molecule tracking system. (**b**) Key steps in SPT, (**c**) Single molecule tracks of Atto647N molecule. (**d**) Intensity trace of a typical single molecule. The plot is taken from, Shangguo Hou, Jack Exell and Kevin Welsher, Real-time 3D single molecule tracking, Nat. Comm. 11, 3607 (2020)

12.2 Single Molecule Force Spectroscopy

Single molecule force spectroscopy has emerged as a unique technique to quantify the forces and motions associated with biomolecules. Of all, the most common force spectroscopy techniques are optical tweezer and AFM. In this section, we will discuss both force spectroscopy techniques and access the range of forces these techniques can measure.

In single-molecule force spectroscopy techniques, a micrometer-sized probe is used to apply force and measure displacement. Since these probes are prone to thermal fluctuations, they are bound to have fundamental resolution limits on displacement and force. In general, these probes (optically trapped bead or AFM cantilever tip) can be approximated as a linear spring with stiffness, k. The thermal noise of the probe determines the spatial resolution, given by,

$$\Delta x = \sqrt{k_B T / k} \tag{12.1}$$

where, Δx is the position noise, $k_B T$ is the thermal energy, and k is the spring stiffness.

From Hook's law ($F = -kx$), the force resolution is given by,

$$\Delta F = k\Delta x = \sqrt{k\, k_B T} \tag{12.2}$$

Here, the magnitude of the noise can be determined by the power spectrum of thermal noise [15] i.e.,

$$S(\nu) = \frac{k_B T}{\pi^2 \beta (\nu^2 + \nu_0^2)} \tag{12.3}$$

where, ν is the frequency, β is the drag on the probe (for example, a sphere of radius a gets a drag of $\beta = 6\pi\eta a$ in a medium of viscosity η), and $\nu_0 = k/(2\pi\beta)$ is the characteristic frequency of the motion.

Note that the area under the power-spectrum gives the positional variance, $(\Delta x)^2$ (see, Fig. 12.2). Now if the cut-off frequency ($\Delta\nu$) is much less than roll-off frequency (ν_0) of the power-spectrum, the positional noise amplitude is given by [15],

$$\Delta x = \sqrt{4\beta k_B T \Delta\nu / k^2} \tag{12.4}$$

and the force resolution becomes,

$$\Delta F = \sqrt{4\beta k_B T \Delta\nu} \tag{12.5}$$

The above equations represents the fundamental resolution limits in force spectroscopy.

In general, the position and force calibrations for force spectroscopy techniques (optical tweezer, magnetic tweezer, and atomic force microscopy) are similar. To elaborate, we take an example of an optical tweezer, which is widely used for force spectroscopy. The probe is treated as a linear spring, the Brownian motion of which is related to spring stiffness, and the force is determined using Hooke's law i.e, $F = -kx$. The position is calibrated by moving the probe by a known distance while recording the position signal. The determination of displacement and force requires calibrating the position detector and measuring the stiffness of the optical trap. The calibration is carried out by moving the trapped bead by a known distance while recording the position signal. This is accomplished by moving a bead stuck to the surface using a piezoelectric stage [16]. The fact that, Brownian motion of a microscopic particle in a trap changes (due to constrained imposed by optical force) is essential to estimate trap stiffness. A particle in a trap is equivalent to the Brownian motion of an over-damped spring caused by the thermal fluctuations of the trapped bead. Here, we make use of equipartition theorem which states that each degree of freedom has a thermal energy of,

$$E - \frac{1}{2} k_B\, T \tag{12.6}$$

where, k_B is the Boltzmann constant and T is the temperature.

Now, equating this to the energy associated with the fluctuations of the trapped particle, we get,

$$\frac{1}{2} k \langle x^2 \rangle = \frac{1}{2} k_B T$$
$$\Rightarrow k = k_B T \langle x^2 \rangle^{-1} \tag{12.7}$$

The above equation states that, the trap-stiffness can be determined if the positional variance of the particle is known.

Another easy method for calculating trap-stiffness is the use of power-spectrum. Noting that, the power spectrum of the fluctuation is a Lorenzian with roll-off frequency, $\nu_0 = k/2\pi\beta$, where, k is the trap-stiffness, and β is the hydrodynamic drag on the bead. The roll-off frequency can be directly obtained by fitting the power spectrum and the drag (depends on particle shape). Figure 12.2 shows a typical power spectrum of a 1 μm particle (attached to a spring undergoing Brownian motion) showing a roll-off frequency of 2 kHz (red) [15]. The disadvantage of this method is however the knowledge of drag coefficient and viscosity which may be challenging to obtain for living systems such as cell.

Since both the above calibration methods use thermal fluctuations, they provide thermal fluctuations near the equilibrium position of the trap. To determine trap-stiffness at larger displacements and determining the region over which the stiffness is constant requires the use of external force to displace the trapped particle. This can be accomplished by moving the trapping chamber placed on a piezoelectric stage and calculating the drag force to measure the trap-stiffness.

Another recent development is the invention of planar lightsheet optical trap (LOT) for confining objects to a single selective plane, rather than a point [17, 18]. This technique use a tightly-focused diffraction-limited sheet of light to trap various objects such as, dielectric

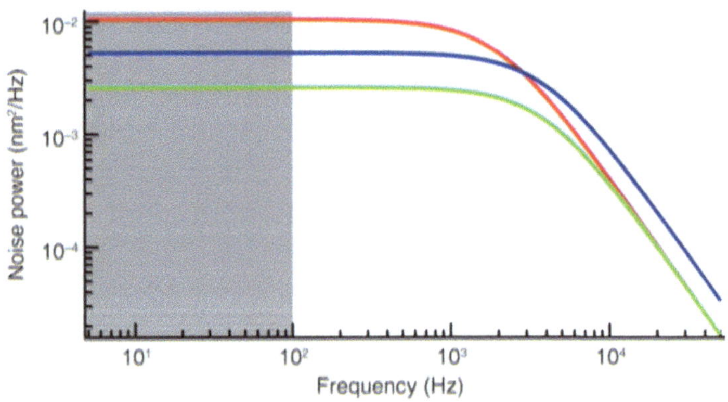

Fig. 12.2 Power spectrum of a 1 μm size particle (red curve). Blue and green curves are for the same particle with half the drag coefficient and for a twofold increase in stiffness (green). The plot is taken from, Keir C Neuman and Attila Nagy, Nature Meth. 5, 491–505 (2008)

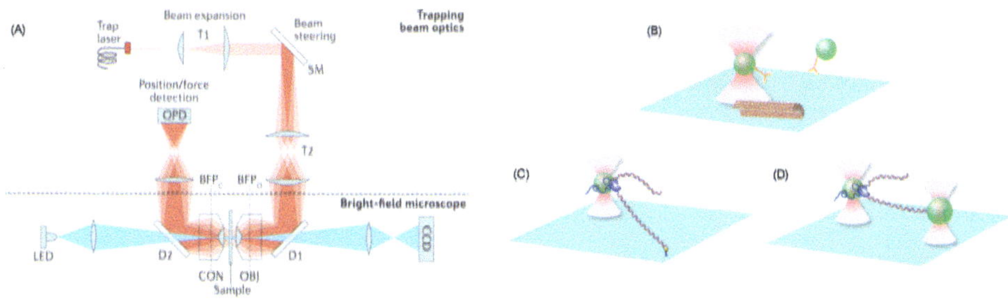

Fig. 12.3 (**a**) Optical configuration of a typical optical trap. (**b–d**) Schematics of assays used in optical tweezers. The figure is taken from, C. J. Bustamante et al., Nature Rev. Meth. Prim. 1, 25 (2021); and, Neuman et al., Nature Meth. 5, 491–505 (2008)

beads and live cells. The planar trap is expected to facilitate new kinds of experiments (confined to 2D) in natural sciences (2D physics, 2D Biology) and engineering.

Prominent applications of optical tweezers are many. For example, the assay shown in Fig. 12.3b enables the determination of force and displacement generated by the individual kinesin molecule as it walks on the microtubule from the displacement of the bead in the optical trap [19]. In another assay (also called, tethered assay) as shown in Fig. 12.3c, long transcriptional records (at a constant force) can be obtained by moving the stage to compensate for this motion (the bead is pulled along the DNA by the polymerase as the DNA is transcribed), by keeping the bead at the same position in the optical trap [20]. Finally, the dumbell assay (Fig. 12.3d) in which, both the ends of DNA are attached to two different optical tweezers. This is similar to that of tethered assay. Here, one of the trap is moved ensuring that the force on the bead is kept constant [1]. There are several such studies that use optical trap for delicate force measurements.

12.2.1 Atomic Force Microscopy

Another popular technique for force measurements is the atomic force microscopy [21,22]. This is conceptually simple and a straightforward technique. Note that, AFM is essentially an imaging tool to map out surfaces with nanometer resolution. Moreover, the technique allows precision measurements of inter- and intra- molecular forces in the range of pico-Newton. However, the technique has earned a new name "Molecular Force Probe (MFP)" when used for one-dimensional force measurements. Specifically, in MFP applications, the cantilever is moved only in the vertical direction, perpendicular to the specimen plane [23,24]

The system works by ensuring that the protein-of-interest is anchored between the cantilever tip and surface of a coverslip (see, Fig. 12.4a). Here, the protein is assumed equivalent to a mechanical spring that exert a restoring force when stretched. The process begins by lowering the tip to the surface using the piezoelectric actuators. After initial

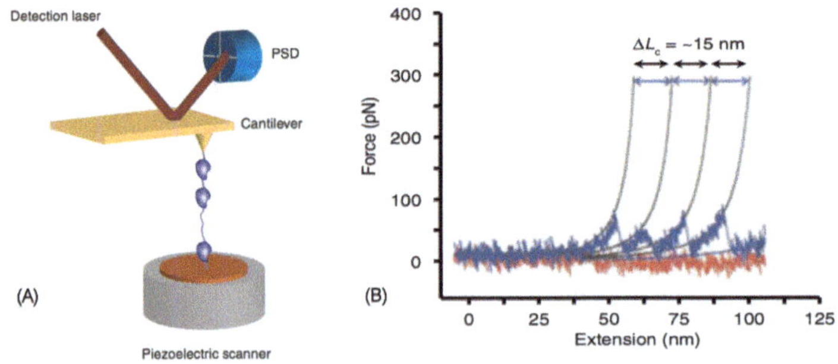

Fig. 12.4 (**a**) Schematic of a typical AFM used for force measurements displaying a pulling experiment of a polyprotein molecule (purple) attached to the tip and sample surface. (**b**) Force-induced unfolding of individual domain repeats of a hypothetical tetramer protein as displayed by MFP stretching curve. The figure is taken from, Keir C Neuman et al., Nature Meth. 5, 491–505 (2008)

contact, the tip is touched to the sample surface. This is followed by retracting the piezoelectric sample stage along the axial/vertical direction forcing the protein/RNA/DNA on the tip to stretch thereby causing the cantilever to bend. The deflection of the cantilever is measured from the deflection of laser spot on a position-sensitive detector (PSD). The force is calculated from the bending of cantilever (of known spring constant). The value of this force is obtained by Hooke's Law ($F = -kx$, where, k is the spring stiffness, and x is the displacement). Typical stiffness values are in the range, 10^1–10^5 pN/nm, and the extension is in the range of few tens of nanometers. To take care of drift and hysteresis associated with piezoelectric actuators, close-loop position feedback control is used. A typical example of hypothetical tetramer protein is shown in Fig. 12.4b. The MFP stretching curve shows the unfolding of individual domain repeats of the protein. Worm-like chain WLC fit to the data gives a persistence length of $\sim$0.8 nm, and a length increment of, $\sim$15 nm [25]. However, the problem of pulling a single versus an ensemble remains and may be addressed in several ways. One way is to ensure a specific attachment to a single molecule by employing a extremely sharp functionalized carbon nanotubes [26].

Optical tweezer and AFM-based force spectroscopy techniques have enabled several delicate force experiments that are otherwise not possible by any other means. These force techniques are continuously evolving and helping our understanding of forces at the single molecule level.

12.3 Event based Single Molecule Localization Microscopy

Within the ambit of SMLM, most of the techniques provide resolutions far beyond classical diffraction limit, and some even down to sub-10 nm. Another way of looking at SMLM is to see fluorescence signal from single molecule as an event. This ensures selective recognition of single molecules from the background. Event-based detection however requires differential sensing i.e., change of intensity/photons with respect to previous measurement by the image sensor. In this section, we will learn how an event image sensor works, and how this enables single molecule imaging.

As compared to the standard frame-based cameras that record absolute brightness, event-based cameras respond to changes in brightness (log intensity) independently for each pixel. Each pixel in the image sensor memorizes the log intensity at each time and continuously monitors for a change. The camera senses an event for a change exceeding the threshold, upon which the location (x, y), the time (t), and the polarity (p) are recorded. Here, 1-bit polarity represents either increased brightness $(+1)$ or decreased brightness (-1). As far as the speed is concerned, an event camera can achieve readout rates of upto 2 MHz [27].

The pixels in an event camera are designed to record and respond independently. They respond to changes in their photo-currents, $I_{pc} = log(I)$, which corresponds to brightness. An event $e_k = (r_k, t_k, p_k)$ triggered at a pixel $r_k = (x_k, y_k)$ (at location, r_k) at time t_k when the change in brightness since the last event i.e., $\Delta I_{pc}(r_k, t_k) = I_{pc}(r_k, t_k) - I_{pc}(r_k, t_k - \Delta t_k)$ reaches a temporal contrast threshold (say, $\pm T$, with $T > 0$) i.e,

$$\Delta I_{pc}(r_k, t_k) = p_k T \tag{12.8}$$

where, $p_k = +1, -1$ is the polarity indicating increasing and decreasing brightness. Note that Δt_k is the time elapsed since the last event at the pixel, r_k, and the pixel bias current determines the contrast sensitivity T. Typical thresholds are set between 15–50%, where the lower limit on T is determined by noise (short noise, thermal noise, and circuit noise).

During data acquisition, the camera sets a threshold on the change in brightness since the last event. An event is described as a change in brightness with time and can be interpreted as a temporal derivative of brightness. For small Δt_k, the increment, $\Delta I_{pc}(r_k, t_k)$ can be approximated by Taylor series expansion i.e.,

$$\begin{aligned} \Delta I_{pc}(r, t) &= I_{pc}(r, t) - I_{pc}(r, t - \Delta t) \\ &\approx \frac{\partial I_{pc}}{\partial t}(r, t)\, \Delta t \end{aligned} \tag{12.9}$$

Neglecting higher orders, the expression for an event is given by,

$$\frac{\partial I_{pc}}{\partial t}(r_k, t_k) \approx \frac{p_k T}{\Delta t_k} \tag{12.10}$$

The above equation summarizes the working of a differential detector such as event/neuromorphic camera.

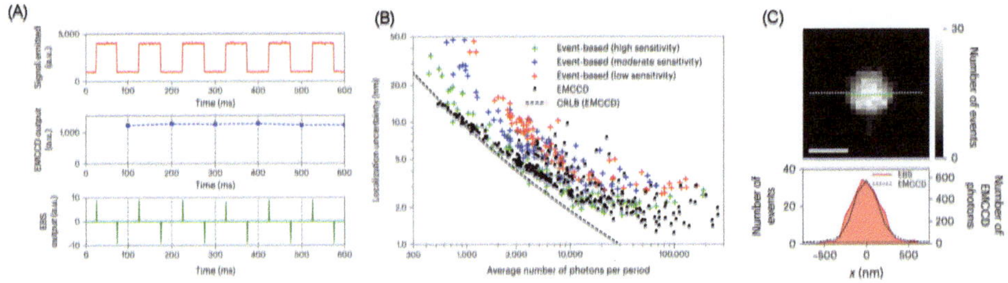

Fig. 12.5 (**a**) Response of emitted signal versus the response of EMCCD and event camera. This shows that event camera responds only during the change of signal intensity i.,e rise and fall of the signal. (**b**) Localization precision versus number of photons per period for event and EMCCD camera. Cramer-Rao lower bound is also shown for EMCCD. (**c**) the number of photons (for EMCCD) and number of events per PSF per period (event). Images are adapted from, Nature Photonics 17, 105–1113 (2023)

12.3.1 Event Based Detection for Standard SMLM Imaging

In this approach, event based detection is carried out to reconstruct the map of single molecule distribution in a specimen [11]. General characteristics of event camera specific to SMLM is shown in Fig. 12.5. The localization density, the calculation of position and localization precision changes for event based SMLM as discussed below:

Standard Density Localization In the standard scheme, the single molecule data is recorded on the event camera. The localization of single molecules is similar to that of PALM/fPALM/STORM. The events are binned into frames with a given time interval $\Delta t = 30\,\text{ms}$ that is generally used for PSF detection. Each molecule is localized using a subset of events which correspond to a spatial ROI (say, ± 4 pixels) about the center of PSF. Then the position x, y is determined by calculating the center-of-mass of all the events or by Gaussian fitting. The number of photons are determined and standard Thompson formula is used for localizing the molecules. A photon number of 5000 is reported for rhodamine or cyanine dyes (Alexa Fluor 647 (AF647) [11].

High Density Localization For this case, the event data is resampled at a shorter timescales. The frames are generated for positive events only with a time bin of $\Delta t = 10\,\text{ms}$, and the localization is performed. This gives the exact time when the molecule turns ON, while the neighboring molecule maintains a constant level of emission. The same is carried out for negative events to extract the moment when the molecule turns OFF. So, each rising edge generates a positive localization, and each falling edge generates a negative localization. Then the molecule is counted, and rest follows similar to that of standard density localization to determine the position and localization precision. To demonstrate, high-density DNA-PAINT experiments on fixed COS-7 cells with DNA-PAINT Atto655 fluorophore labelling against α-tubulin is shown in Fig. 12.6.

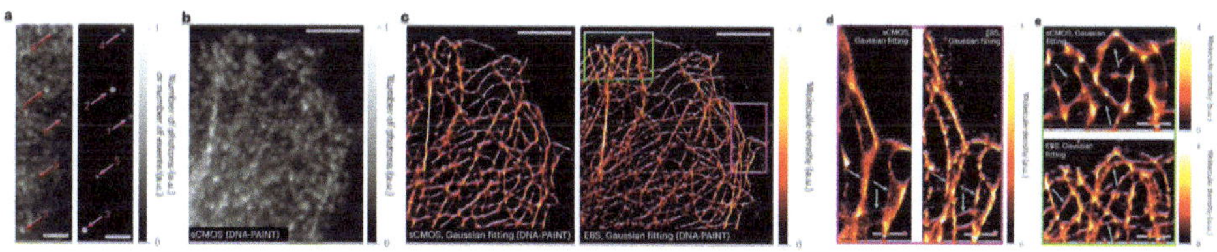

Fig. 12.6 (**a**) Images of dense 488 nm and sparse 647 nm fluorescent beads obtained using EMCCD (left), and event (right). (**b**) Raw sCMOS frame (50 ms exposure) showing the extreme density of molecules. Alongside SMLM images with sCMOS (**c**. left) and event-based (**c**. right) are also shown. (**d**, **e**) Enlarged view of two different regions. Images are adapted from, Nature Photonics 17, 105–1113 (2023)

12.3.2 Event Based Detection for High Spatio-Temporal 3D SMLM Imaging

This technique (termed as $e3DSMLM$) takes the event detection to a new level that exploits the changes in the number of photons during the blinking time [12]. In a way, the technique sub-samples the PSF into several temporal PSFs ($tPSFs$) and determines the position and localization precision of the molecule. This is possible due to the near continuous mode of event-based data collection with negligible read time. For example, the detector allows an exposure of 3 ms with a read time of $\sim$100 μs.

In this approach, the data of single molecule PSFs are recorded at a lower time period than the blinking time period of the molecule. It was observed that, single molecules emit intermittently during the blinking period as seen in Fig. 12.7a. So, recording PSFs multiple times in a single blinking period gives several realizations of the single molecule, thereby locating its position with some uncertainty and a center. As an example, consider recording of sparsely spread $Cy3$ single molecules on a coverslip at sampling rates of, 333 Hz (event

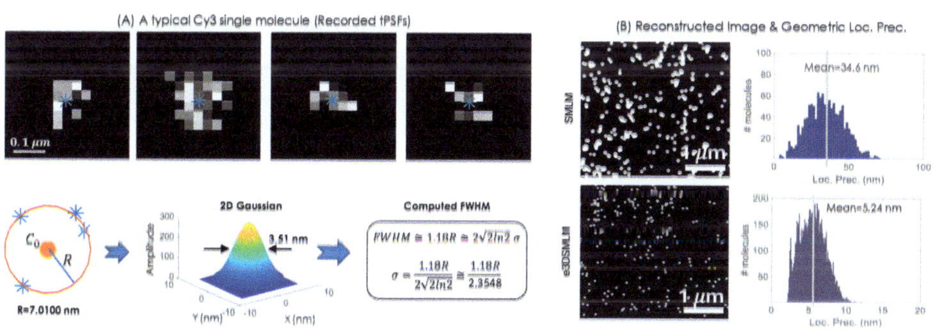

Fig. 12.7 Recorded tPSFs of a typical single Cy3 molecule for a collection time of 100 ms. The corresponding fit-circle followed by a 2D Gaussian function determine its location and geometric localization precision. Alongside, reconstructed SMLM and $e3DSMLM$ reconstructed images with localization precision plots are also shown. Image taken from, J. Basumatary et al., BioRxiv, 2024. DOI: https://doi.org/10.1101/2023.12.30.573392

collection time of $\tau = 3\,\text{ms}$), 100 Hz ($\tau = 10\,\text{ms}$) and 33 Hz ($\tau = 30\,\text{ms}$). The localization precision of the detected molecules are calculated using geometric localization method. In this method, each molecule is rendered as a 2D Gaussian spot having width equals to calculated localization precession and intensity A is the sum of all the events (both positive and negative). A circle is fit through centroids of segmented PSF giving the fitting parameters, $r,\ a,\ b$ from which the center $C_0(a, b)$ and radius r are determined (see, Fig. 12.7). Subsequently, a Gaussian is used to render PSF with its center as the position of the molecule and l_p as the localization precision i.e,

$$PSF \propto \exp\left[-\frac{(x-a)^2+(y-b)^2}{2l_p^2}\right] \tag{12.11}$$

Following Refs. [28, 29], the relation between r of a circle and FWHM of a Gaussian is,

$$FWHM \approx 1.18\, r \tag{12.12}$$

Noting that l_p is related to FWHM of the Gaussian i.e, $FWHM = 2\sqrt{2\,ln2}\,l_p$, we get,

$$1.18\, r = 2\sqrt{2\,ln2}\,l_p \quad \Rightarrow l_p \approx \frac{1.18\,R}{2.355} \tag{12.13}$$

The above equation is used to calculate the localization precision of single molecules geometrically.

The event-based reconstructed super-resolved image of Cy3 molecules is shown Fig. 12.7b. The localization precision of these molecules is calculated using geometric localization method (see, Fig. 12.7a). For comparison, reconstruction for standard SMLM (data acquired by EMCCD) along with localization precision are also shown. Alongside blinking characteristics and images (as seen by neuromorphic camera) of a single molecule is also shown. Fig. 12.7a shows the distribution of a typical single-molecule. Since the blinking period of a molecule is much longer (in tens of milliseconds), the event camera captures multiple realizations of the bright PSF spots. The recorded $tPSFs$ is also shown in Fig. 12.7a. It is apparent that the photon emission from a single molecule is random, and the corresponding $tPSFs$ have distinct distributions. This is important from the point-of-view of super-resolution imaging since this allows the localization of single molecules purely based on geometric parameters estimated from multiple $tPSFs$. Subsequently, centroid of the spot is determined from consecutive frames (see, blue star in Fig. 12.7a). Assuming symmetric localization precision, a circle is fit (with a goodness-of-fit value, ~0.9) to the centroid points (red circle) and the center of the circle (orange dot) is determined. A Gaussian corresponding to the circle is identified and FWHM is calculated. This gives the position and localization precision (FWHM) for a single molecule. The corresponding localization parameters for a few Cy3 single molecules are tabulated in

Fig. 12.7b. The same process is repeated for all the molecules and a super-resolved map is constructed. The localization precision plots indicate an average localization of 4.581 nm (equivalent to a resolution of 10.53 nm) at $\tau = 3$ ms for a typical $Cy3$ molecule. This is about 7 times less than the localization precision of standard SMLM that uses well-known Thompson's formula [36]. The corresponding spatial resolution is approximately, $R = 2.3l_p$, where l_p is the localization precision [37]. Overall, this suggests a spatio-temporal particle resolution of $2.3l_p \times 3\tau = 2.3 \times 4.581 \times 9\,\text{nm} \cdot \text{msec} = 9.48$ par. Note that, $t = 3\tau$ is necessary to obtain the minimum information (atleast 3 centroids to represent a circle) for localizing the molecule. This is way better than standard SMLM.

In another experiment, mitochondrial in a transfected cell is imaged as shown in Fig. 12.8. The epifluorescence image of the mEos-Tom20 transfected cell of which a small region is chosen for study is also shown. Fig. 12.8b shows the reconstructed super-resolved image, and compares it with the reconstructed SMLM image (reconstructed from the data/images captured by EMCCD camera) (see, Fig. 12.8c). Three sub-regions (R1, R2, and R3) are chosen and compared as shown in Fig. 12.8b, c. The corresponding localization plot gives an average localization precision of 5.677 nm for event-based reconstructed image. This is way below the localization precision of standard SMLM which is 40.03 nm, and equivalent to a spatial resolution of, 13.05 nm and 92.07 nm for $eventSMLM$ and EMCCD-based $SMLM$, respectively giving a spatio-temporal resolution of 11.75 par. In this case, the data is obtained at $\tau = 10$ ms.

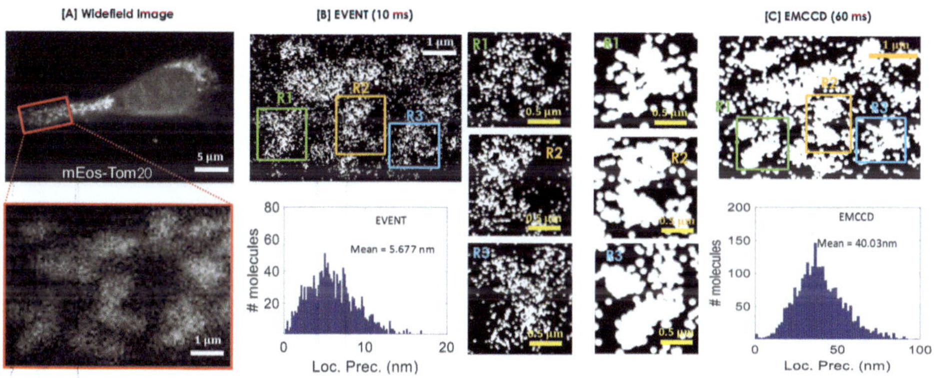

Fig. 12.8 (**a**) Widefield fluorescence image of mEos-Tom20 transfected NIH3T3 cell (fixed). The red zone is chosen as a region of interest. (**b**) Event detection-based reconstructed super-resolved image of Tom20 single molecules. Frames are constructed by collecting events at a sampling time of 10 ms. (**c**) SMLM reconstructed map of Tom20 molecules detected using EMCCD detector (30 ms exposure time). Enlarge views of three chosen regions R1, R2, and R3 from (**b**) and (**c**) are also shown, along with respective localization histograms. Image taken from, J. Basumatary et al., BioRxiv, 2024. DOI: https://doi.org/10.1101/2023.12.30.573392

12.4 Limitations, Applications and Future Prospects

Single-molecule techniques, including measurement and imaging, are key to the new developments in fields ranging from fundamental biology to optical science. This is evident from the rapidly growing number of applications in diverse disciplines of natural sciences. The potential of these techniques can be accessed from their regular use and large-scale commercialization.

These techniques open up the door to new science and lay the foundation to address unsolvable problems and bigger challenges. Among others, one of the challenges is to develop single-molecule techniques that can be used to study living cells and tissues. The technique is still in its early stages where it can only be applied to measurements of purified proteins and other biomolecules. Most of these techniques require fluorescence labeling to carry out the study. In this respect, label-free imaging and measurement techniques (such as, Raman spectroscopy) are the way forward [30]. However, these techniques are immature and laced with several limitations, making them inaccessible to non-specialists.

Although optical tweezers and AFM hold a lot of promise, these techniques have their limitations too. One of the limitations of optical tweezers is the local heating caused by the trapping laser and the nature of the optical trap that depends on the gradient of the optical trap. In addition, optical traps lack selectivity, meaning any dielectric particle can be trapped, and the number of such particles trapped simultaneously can be huge. On the bright side, new advances have diversified optical tweezers and pushed them toward high resolution. A recent development has been the realization of two-dimensional light sheet tweezer (2D-LOT) that can trap particles (dielectric beads and live cells) on a selective single plane over a near-uniform trap region, and enable imaging on the go [18]. Along similar lines, AFM suffers from the large dimension and high stiffness of cantilevers that determine lower bound on force range and bandwidth, which make it difficult to measure forces associated with different biological structures. On the bright side, recently developed high-speed AFMs allow the study of molecular events in real-time [31]. In the future, a combination of one or more techniques (force mapping and spectroscopy) with SMLM may be beneficial for following protein unfolding or binding.

Single-molecule techniques suited for in-vivo experiments are expected to provide a real-time view of the functioning of individual proteins and other biomolecules in their native environment. This will reveal the delicate force needed to carry out these sub-cellular processes to function for supporting life systems. Future single-molecule techniques are required to overcome several difficulties in looking into the interior functioning of a cell without causing much perturbation (less phototoxicity and non-invasive manipulation). In addition, these techniques need to become more sophisticated for studying multi-enzyme complexes and macromolecular machines [32–34]. In this respect, a combination of single-molecule imaging and force spectroscopy techniques seems to be a good recipe for further growth in our understanding of life processes, be it at a cellular or multicellular organism level.

References

1. Abbondanzieri EA et al (2005) Direct observation of base-pair stepping by RNA polymerase. Nature 438:460–465
2. Grandbois M et al (1999) How strong is a covalent bond? Science 283:1727–1730
3. Hohng S et al (2007) Fluorescence-force spectroscopy maps two-dimensional reaction landscape of the holliday junction. Science 318:279–283
4. Presman DM, Ball DA, Paakinaho V, Grimm JB, Lavis LD et al (2017) Quantifying transcription factor binding dynamics at the single-molecule level in live cells. Methods 123:76–88
5. Suzuki KGN, Kasai RS, Hirosawa KM, Nemoto YL, Ishibashi M et al (2012) Transient GPI-anchored protein homodimers are units for raft organization and function. Nat Chem Biol 8:774–783
6. Chen Z, Cao Y, Lin C-W, Alvarez S, Oh D et al (2021) Nanopore-mediated protein delivery enabling three-color single-molecule tracking in living cells. PNAS 118:e2012229118
7. Yildiz A, Forkey JN, McKinney SA, Ha T, Goldman YE, Selvin PR (2003) Myosin V walks hand-over-hand: single fluorophore imaging with 1.5-nm localization. Science 300:2061–2065
8. Lowe AR, Siegel JJ, Kalab P, Siu M, Weis K, Liphardt JT (2010) Selectivity mechanism of the nuclear pore complex characterized by single cargo tracking. Nature 467:600–603
9. Fujiwara T, Ritchie K, Murakoshi H, Jacobson K, Kusumi A (2002) Phospholipids undergo hop diffusion in compartmentalized cell membrane. J Cell Biol 157:1071–1082
10. Li CW, Lim SO, Chung EM, Kim YS, Park AH et al (2018) Eradication of triple-negative breast cancer cells by targeting glycosylated PD-L1. Cancer Cell 33:187–201
11. Cabriel C, Monfort T, Specht CG, Izeddin I (2023) Event-based vision sensor for fast and dense single-molecule localization microscopy. Nat Photon 17:1105–1113
12. Basumatary J, Aravinth S, Pant N, Ramanathan V, Thakur CS, Mondal PP (2023) Event-based single molecule localization microscopy (eventSMLM) for high spatio-temporal super-resolution imaging. https://doi.org/10.1101/2023.12.30.573392
13. Hou S, Exell J, Welsher K (2020) Real-time 3D single molecule tracking. Nat Commun 11:3607
14. Crocker JC, Grier DG (1996) Methods of digital video microscopy for colloidal studies. J Colloid Interface Sci 179:298–310
15. Neuman KC, Nagy A (2008) Single-molecule force spectroscopy: optical tweezers, magnetic tweezers and atomic force microscopy. Nature Methods 5:491–505
16. Vermeulen KC et al (2006) Calibrating bead displacements in optical tweezers using acousto-optic deflectors. Rev Sci Instrum 77:013704
17. Mondal PP et al (2022) Lightsheet optical tweezer (LOT) for optical manipulation of microscopic particles and live cells. Sci Rep 12:10229
18. Baro N, Mondal PP (2024) Planar optical tweezer trap (2D-LOT) system realized by light sheet illumination and orthogonal widefield detection. https://doi.org/10.1101/2024.04.15.589441
19. Block SM et al (1990) Bead movement by single kinesin molecules studied with optical tweezers. Nature 348:348–352
20. Neuman KC et al (2003) Ubiquitous transcriptional pausing is independent of RNA polymerase backtracking. Cell 115:437–447
21. Binnig G, Quate CF, Gerber C (1986) Atomic force microscope. Phys Rev Lett 56:930–933
22. Lee GU, Chrisey LA, Colton RJ (1994) Direct measurement of the forces between complementary strands of DNA. Science 266:771–773
23. Rief M et al (1997) Reversible unfolding of individual titin immunoglobulin domains by AFM. Science 276:1109–1112
24. Rief M et al (1997) Single molecule force spectroscopy on polysaccharides by atomic force microscopy. Science 275:1295–1297

25. Bustamante C, Marko JF, Siggia ED, Smith S (1994) Entropic elasticity of lambda-phage DNA. Science 265:1599–1600
26. Katz E, Willner I (2004) Biomolecule-functionalized carbon nanotubes: applications in nanobioelectronics. ChemPhysChem 5:1084–1104
27. Lichtsteiner P, Posch C, Delbruck T (2008) A 128 × 128 120 dB 15 μs latency asynchronous temporal contrast vision sensor. IEEE J Solid-State Circuits 43(2):566–576
28. Hecht E (2008) Optics, Chapter 13, 4th edn. Pearson
29. https://support.zemax.com/hc/en-us/articles/1500005488161-How-to-convert-FWHM-measurements-to-1-e-2-halfwidths
30. Stepula E et al (2024) Label-free 3D molecular imaging of living tissues using Raman spectral projection tomography. Nat Commun 15:7717
31. Ando T et al (2001) A high-speed atomic force microscope for studying biological macromolecules. Proc Natl Acad Sci USA 98:12468–12472
32. Wen JD et al (2008) Following translation by single ribosomes one codon at a time. Nature 452:598–603
33. van Oijen AM (2007) Single-molecule studies of complex systems: the replisome. Mol Biosyst 3:117–125
34. Lee JB et al (2006) DNA primase acts as a molecular brake in DNA replication. Nature 439:621–624

Index

P. P. Mondal, S. Hess, *Single-Molecule Super-Resolution Optical Microscopy*, https://doi.org/10.1007/978-3-032-13442-4

MIX
Papier aus verantwortungsvollen Quellen
Paper from responsible sources
FSC® C105338

If you have any concerns about our products,
you can contact us on
ProductSafety@springernature.com

In case Publisher is established outside the EU,
the EU authorized representative is:
Springer Nature Customer Service Center GmbH
Europaplatz 3, 69115 Heidelberg, Germany

Printed by Libri Plureos GmbH
in Hamburg, Germany